D0053425

Plastic

Books by Susan Freinkel

*American Chestnut: The Life, Death,
and Rebirth of a Perfect Tree*

Plastic: A Toxic Love Story

Susan Freinkel

Plastic

A Toxic Love Story

Houghton Mifflin Harcourt
BOSTON • NEW YORK
2011

Copyright © 2011 by Susan Freinkel

For information about permission to reproduce selections from this book, write to Permissions, Houghton Mifflin Harcourt Publishing Company, 215 Park Avenue South, New York, New York 10003.

www.hmhbooks.com

Library of Congress Cataloging-in-Publication Data
Freinkel, Susan, date.
Plastic : a toxic love story / Susan Freinkel.
p. cm.
Includes bibliographical references and index.
ISBN 978-0-547-15240-0
1. Plastics. I. Title.
TP1120.F74 2011
620.1'923 — dc22
2010043019

Printed in the United States of America

DOC 10 9 8 7 6 5 4 3 2 1

Excerpt from "Plastic" from *Unincorporated Persons in the Late Hondo Dynasty* by Tony Hoagland. Copyright © 2010 by Tony Hoagland. Reprinted with the permission of Graywolf Press, Minneapolis, Minnesota, www.graywolfpress.org.

For Eli, Isaac, and Moriah

I wonder if it would have done any good then
If I had walked over and explained a few things to them

About Plastic?
About how it is so much easier to stretch than
                                    human nature,

which accounts for some of the strain imposed on
                                    the late 20th-century self . . .

                    —Tony Hoagland, "Plastic"

Contents

Plastic

Plasticville

I N 1950, a Philadelphia toy company came out with a new accessory for electric-train enthusiasts: snap-together kits of plastic buildings for a place it called Plasticville, U.S.A. Sets of plastic people to populate the town were optional.

It started as a sleepy, rural place where trains might roll past red-sided barns to pull into a village with snug Cape Cod homes, a police department, a fire station, a schoolhouse, and a quaint white church with a steeple. But over the years, the product line spread into a bustling burb of housing tracts filled with two-story Colonials and split-level ranch houses and a Main Street that boasted a bank, a combination hardware store/pharmacy, a modern supermarket, a two-story hospital, and a town hall modeled on Philadelphia's historic Independence Hall. Eventually Plasticville even gained a drive-in motel, an airport, and its own TV station, WPLA.

Today, of course, we all live in Plasticville. But it wasn't clear to me just how plastic my world had become until I decided to go an entire day without touching anything plastic. The absurdity of this experiment became apparent about ten seconds into the appointed morning when I shuffled bleary-eyed into the bathroom: the toilet seat was plastic. I quickly revised my plan. I would spend the day writing down everything I touched that was plastic.

Within forty-five minutes I had filled an entire page in my Penway Composition Book (which itself had to be cataloged as partly plastic, given its synthetic binding, as did my well-sharpened no. 2 pencil, which was coated with yellow paint that contained acrylic). Here's some of what I wrote down as I made my way through my early-morning routine:

Alarm clock, mattress, heating pad, eyeglasses, toilet seat, tooth-brush, toothpaste tube and cap, wallpaper, Corian counter, light switch, tablecloth, Cuisinart, electric teakettle, refrigerator handle, bag of frozen strawberries, scissors handle, yogurt container, lid for can of honey, juice pitcher, milk bottle, seltzer bottle, lid of cin-namon jar, bread bag, cellophane wrapping of box of tea, packag-ing of tea bag, thermos, spatula handle, bottle of dish soap, bowl, cutting board, baggies, computer, fleece sweatshirt, sports bra, yoga pants, sneakers, tub containing cat food, cup inside tub to scoop out the kibble, dog leash, Walkman, newspaper bag, stray packet of mayo on sidewalk, garbage can.

"Wow!" said my daughter, her eyes widening as she scanned the rap-idly growing list.

By the end of the day I had filled four pages in my notebook. My rule was to record each item just once, even those I touched repeat-edly, like the fridge handle. Otherwise I could have filled the whole notebook. As it was, the list included 196 entries, ranging from large items, like the dashboard of my minivan—really, the entire inte-rior—to minutiae, like the oval stickers adorning the apples I cut up for lunch. Packaging, not surprisingly, made up a big part of the list.

I'd never thought of myself as having a particularly plastic-filled life. I live in a house that's nearly a hundred years old. I like natural fabrics, old furniture, food cooked from scratch. I would have said my home harbors less plastic than the average American's—mainly for aesthetic reasons, not political ones. Was I kidding myself? The next day I tracked everything I touched that *wasn't* made of plastic. By bedtime, I had recorded 102 items in my notebook, giving me a plastic/nonplastic ratio of nearly two to one. Here's a sample from the first hour of the day:

> Cotton sheets, wood floor, toilet paper, porcelain tap, strawberries, mango, granite-tile countertop, stainless steel spoon, stainless steel faucet, paper towel, cardboard egg carton, eggs, orange juice, aluminum pie plate, wool rug, glass butter dish, butter, cast-iron griddle, syrup bottle, wooden breadboard, bread, aluminum colander, ceramic plates, glasses, glass doorknob, cotton socks, wooden dining-room table, my dog's metal choke collar, dirt, leaves, twigs, sticks, grass (and if I weren't using a plastic bag, what my dog deposited amid those leaves, twigs, and grass).

Oddly, I found it harder and more boring to maintain the nonplastic list. Because I'd pledged not to count items more than once, after the first flood of entries, there wasn't that much variety — at least not when compared with the plastics catalog. Wood, wool, cotton, glass, stone, metal, food. Distilled further: animal, vegetable, mineral. Those basic categories pretty much encompassed the items on the nonplastic list. The plastic list, by contrast, reflected a cornucopia of materials, a dazzling variety of the synthetica that has come to constitute such a huge, and yet strangely invisible, part of modern life.

Pondering the lengthy list of plastic in my surroundings, I realized I actually knew almost nothing about it. What is plastic, really? Where does it come from? How did my life become so permeated by synthetics without my even trying? Looking over the list I could see plastic products that I appreciated for making my life easier and more convenient (my wash-and-wear clothes, my appliances, that plastic bag for my dog's poop) and plastic things I knew I could just as easily do without (Styrofoam cups, sandwich baggies, my nonstick pan).

I'd never really looked hard at life in Plasticville. But news reports about toxic toys and baby bottles seemed to suggest that the costs might outweigh the benefits. I began to wonder if I'd unwittingly exposed my own children to chemicals that could affect their development and health. That hard-plastic water bottle I'd included in my daughter's lunch since kindergarten has been shown to leach a chemical that mimics estrogen. Was that why she'd sprouted breast buds at nine? Other questions quickly followed. What was happening to the plastic things I diligently dropped into my recycling bin? Were they actually being recycled? Or were my discards ending up far away in the ocean in vast currents of plastic trash? Were there seals somewhere choking on my plastic bottle tops? Should I quit using plastic shopping bags? Would that soda bottle really outlive my children and me? Did it matter? Should I care? What does it really mean to live in Plasticville?

The word *plastic* is itself cause for confusion. We use it in the singular, and indiscriminately, to refer to any artificial material. But there are tens of thousands of different plastics.* And rather than making up a single family of materials, they're more a collection of loosely related clans.

I got a glimpse of the nearly inexhaustible possibilities contained in that one little word when I visited a place in New York called Material ConneXion, a combination of a consultancy and a materials larder for designers pondering what to make their products out of. Its founder described it as a "petting zoo for new materials." And I did feel like I was in a tactile and visual wonderland as I browsed some of the thousands of plastics on file. There was a thick acrylic slab that looked like a pristine frozen waterfall; jewel-colored blobs of gel that begged to be squeezed; a flesh-toned fabric that looked and felt like an old person's skin. ("Ugh, I'd never want to wear anything like that," one staffer commented.) There were swatches of fake fur, green netting, gray shag rug, fake blades of grass, fabric that holds the memory of how it's folded, fabric that can absorb solar energy and transmit it to the wearer. I looked at blocks that mimicked finely

---

* For a brief description of the more common plastics, see "Cast of Characters" at the end of the book.

veined marble, smoky topaz, dull concrete, speckled granite, grained wood. I touched surfaces that were matte, shiny, bumpy, sandpapery, fuzzy, squishy, feathery, cool as metal, warm and yielding as flesh.

But a plastic doesn't have to be part of the exotic menagerie at Material ConneXion to impress. Even a common plastic such as nylon offers wow-inducing possibility. It can be silky when serving in a parachute, stretchy when spun into pantyhose, bristly when fixed at the end of your toothbrush, or bushy on a strip of Velcro. *House Beautiful* swooned over such versatility in a 1947 article titled "Nylon . . . the Gay Deceiver."

However much they differ, all plastics have one thing in common: they are polymers, which is Greek for "many parts." They are substances made up of long chains of thousands of atomic units called monomers (Greek for "one part") linked into giant molecules. Polymer molecules are absurdly huge compared to the tidy, compact molecules of a substance like water, with its paltry one oxygen and two hydrogen atoms. Polymer molecules can contain tens of thousands of monomers — chain links so long that for years scientists disputed whether they could actually be bonded into a single molecule. You might as well claim, said one chemist, that "somewhere in Africa an elephant was found who was 1,500 feet long and 300 feet high." But the molecules did exist, and their hugeness helps account for plastic's essential feature: its plasticity. Think of the ways a long strand of beads can be manipulated — pulled or stretched, stacked or coiled — compared to what can be done with just a single bead or a few. The lengths and arrangement of the strands help to determine a polymer's properties: its strength, durability, clarity, flexibility, elasticity. Chains crowded close together can make for a tough, rigid plastic bottle, like the kind used to hold detergent. Chains more widely spaced can yield a more flexible bottle ideal for squeezing out ketchup.

It's often said that we live in the age of plastics. But when, exactly, did we slip into that epoch? Some say it began in the mid-nineteenth century, when inventors started developing new, malleable semisynthetic compounds from plants to replace scarce natural materials such as ivory. Others fix the date to 1907, when Belgian émigré Leo Baekeland cooked up Bakelite, the first fully synthetic polymer, made entirely of molecules that couldn't be found in nature. With the product's invention, the Bakelite Corporation boasted, humans had transcended the classic taxonomies of the natural world: the animal, mineral, and vegetable kingdoms. Now we had "a fourth kingdom, whose boundaries are unlimited."

You could also peg the dawn of the plastics age to 1941, when, shortly after the bombing of Pearl Harbor, the director of the board responsible for provisioning the American military advocated the substitution, whenever possible, of plastics for aluminum, brass, and other strategic metals. World War II pulled polymer chemistry out of the lab and into real life. Many of the major plastics we know today—polyethylene, nylon, acrylic, Styrofoam—got their first marching orders during the war. And having ramped up production to meet military needs, industry inevitably had to turn its synthetic swords into plastic plowshares. As one early plastics executive recalled, by the war's end it was obvious that "virtually nothing was made from plastic and anything could be." That's when plastics truly began infiltrating every pore of daily life, quietly entering our homes, our cars, our clothes, our playthings, our workplaces, even our bodies.

In product after product, market after market, plastics challenged traditional materials and won, taking the place of steel in cars, paper and glass in packaging, and wood in furniture. Even Amish buggies are now made partly out of the fiber-reinforced plastic known as fiberglass. By 1979, production of plastics exceeded that of steel. In an astonishingly brief period, plastic had become the skeleton, the connective tissue, and the slippery skin of modern life.

Indisputably, plastic does offer advantages over natural materials. Yet that doesn't fully account for its sudden ubiquity. Plasticville became possible—and perhaps even inevitable—with the rise of the petrochemical industry, the behemoth that came into being in the 1920s and '30s when chemical companies innovating new polymers began to align with the petroleum companies that controlled the essential ingredients for building those polymers.

Oil refineries run 24-7 and are continuously generating byproducts that must be disposed of, such as ethylene gas. Find a use for that gas, and your byproduct becomes a potential economic opportunity. Ethylene gas, as British chemists discovered in the early 1930s, can be made into the polymer polyethylene, which is now widely used in packaging. Another byproduct, propylene, can be redeployed as a feedstock for polypropylene, a plastic used in yogurt cups, microwavable dishes, disposable diapers, and cars. Still another is the chemical acrylonitrile, which can be made into acrylic fiber, making possible that quintessential emblem of our synthetic age AstroTurf.

Plastics are a small piece of the petroleum industry, representing a minor fraction of the fossil fuels we consume. But the economic imperatives of the petroleum industry have powered the rise of Plasticville. As environmentalist Barry Commoner argued: "By its own internal logic, each new petrochemical process generates a powerful tendency to proliferate further products and displace preexisting ones." The continuous flow of oil fueled not just cars but an entire culture based on the consumption of new products made of plastics. This move into Plasticville wasn't a considered decision, the result of some great economic crisis or political debate. Neither did it take into account social good or environmental impact or what we were supposed to do with all our plastic things at the end of their useful lives. Plastic promised abundance on the cheap, and when in human history has that ever been a bad thing? No wonder we became addicted to plastic, or, rather, to the convenience and comfort, safety and security, fun and frivolity that plastic brought.

The amount of plastic the world consumes annually has steadily risen over the past seventy years, from almost nil in 1940 to closing in on six hundred billion pounds today. We became plastic people really just in the space of a single generation. In 1960, the average American consumed about thirty pounds of plastic products. Today, we're each consuming more than three hundred pounds of plastics a year, generating more than three hundred billion dollars in sales. Considering that lightning-quick ascension, one industry expert declared plastics "one of the greatest business stories of the twentieth century."

The rapid proliferation of plastics, the utter pervasiveness of it in our lives, suggests a deep and enduring relationship. But our feelings toward plastic are a complicated mix of dependence and distrust — akin to what an addict feels toward his or her substance of choice. Initially, we reveled in the seeming feats of alchemy by which scientists produced one miraculous material after another out of little more than carbon and water and air. It's "wonderful how du Pont is improving on nature," one woman gushed after visiting the company's Wonder World of Chemistry exhibit at a 1936 Texas fair. A few years later, people told pollsters they considered *cellophane* the third most beautiful word in the English language, right behind *mother* and *memory*. We were prepared, in our infatuation, to believe only the very best of our partner in modernity. Plastics heralded a new era of material freedom, liberation from nature's stinginess. In the plastic age, raw materials would not be in short supply or constrained by their innate properties, such as the rigidity of wood or the reactivity of metal. Synthetics could substitute for, or even precisely imitate, scarce and precious materials. Plastic, admirers predicted, would deliver us into a cleaner, brighter world in which all would enjoy a "universal state of democratic luxury."

It's hard to say when the polymer rapture began to fade, but it was gone by 1967 when the film *The Graduate* came out. Somewhere along the line — aided surely by a flood of products such as pink flamingos, vinyl siding, Corfam shoes — plastic's penchant for inexpensive imitation came to be seen as cheap ersatz. So audiences knew exactly why Benjamin Braddock was so repelled when a family friend took him aside for some helpful career advice: "I just want to say one word to you . . . Plastics!" The word no longer conjured an enticing horizon of possibility but rather a bland, airless future, as phony as Mrs. Robinson's smile.

Today, few other materials we rely on carry such a negative set of associations or stir such visceral disgust. Norman Mailer called it "a malign force loose in the universe . . . the social equivalent of cancer." We may have created plastic, but in some fundamental way it remains essentially alien — ever seen as somehow unnatural (though it's really no less natural than concrete, paper, steel, or any other manufactured material). One reason may have to do with its preternatural endurance. Unlike traditional materials, plastic won't dissolve or rust or break down — at least, not in any useful time frame. Those long polymer chains are built to last, which means that much of the plastic we've produced is with us still — as litter, detritus on the ocean floor, and layers of landfill. Humans could disappear from the earth tomorrow, but many of the plastics we've made will last for centuries.

This book traces the arc of our relationship with plastics, from enraptured embrace to deep disenchantment to the present-day mix of apathy and confusion. It's played out across the most transformative century in humankind's long project to shape the material world to its own ends. The story's canvas is huge but also astonishingly familiar, because it is full of objects we use every day. I have chosen eight to help me tell the story of plastic: the comb, the chair, the Frisbee, the IV bag, the disposable lighter, the grocery bag, the soda bottle, the credit card. Each offers an object lesson on what it means to live in Plasticville, enmeshed in a web of materials that are rightly considered both the miracle and the menace of modern life. Through these objects I examine the history and culture of plastics and how plastic things are made. I look at the politics of plastics and how synthetics are affecting our health and the environment, and I explore efforts to develop more sustainable ways of producing and disposing of plastics. Each object opens a window onto one of Plasticville's many precincts. It is my hope that taken together, they shed light on our relationship with plastic and suggest how, with effort, it might become a healthier one.

Why did I decide to focus on such small, common things? None have the razzle-dazzle that cutting-edge polymer science is delivering, such as smart plastics that can mend themselves and plastics that conduct electricity. But those are not the plastic things that play meaningful roles in our everyday lives. I also chose not to use any durable goods, such as cars or appliances or electronics. No question any of these could have offered insights into the age of plastics. But the material story of a car or an iPhone encompasses far more than just plastics. Simple objects, properly engaged, distill issues to their essence. As historian Robert Friedel notes, it's in the small things "that our material world is made."

Simple objects sometimes tell tangled stories, and the story of plastics is riddled with paradoxes. We enjoy an unprecedented level of material abundance and yet it often feels impoverishing, like digging through a box packed with Styrofoam peanuts and finding nothing else there. We take natural substances created over millions of years, fashion them into products designed for a few minutes' use, and then return them to the planet as litter that we've engineered to never go away. We enjoy plastics-based technologies that can save lives as never before but that also pose insidious threats to human health. We bury in landfills the same kinds of energy-rich molecules that we've scoured the far reaches of the earth to find and excavate. We send plastic waste overseas to become the raw materials for finished products that are sold back to us. We're embroiled in pitched political fights in which plastic's sharpest critics and staunchest defenders make the same case: these materials are too valuable to waste.

These paradoxes contribute to our growing anguish over plastics. Yet I was surprised to discover how many of the plastics-related issues that dominate headlines today had surfaced in earlier decades. Studies that show traces of plastics in human tissue go back to the 1950s. The first report of plastic trash in the ocean was made in the 1960s. Suffolk County, New York, enacted the first ban on plastic packaging in 1988. In every case, the issues seized our attention for a few months or even years and then slipped off the public radar.

But the stakes are much higher now. We've produced nearly as much plastic in the first decade of this millennium as we did in the entire twentieth century. As Plasticville sprawls farther across the landscape, we become more thoroughly entrenched in the way of life it imposes. It is increasingly difficult to believe that this pace of plasticization is sustainable, that the natural world can long endure our ceaseless "improving on nature." But can we start engaging in the problems plastics pose? Is it possible to enter into a relationship with these materials that is safer for us and more sustainable for our offspring? Is there a future for Plasticville?

Improving on Nature

IF YOU GO ON EBAY, that virtual souk of human desire, you'll find a small but dedicated trade in antique combs. Trawling the site on various occasions, I've seen dozens of combs made of the early plastic called celluloid — combs so beautiful they belonged in a museum, so beguiling I coveted them for my own. I've seen combs that looked as if they were carved from ivory or amber, and some that were flecked with mica so they shone as if made of hammered gold. I've seen huge, lacy decorative combs of faux tortoiseshell that might have crowned the piled-high up-twist of a Gilded Age debutante, and tiara-like combs twinkling with sapphire or emerald or jet "brilliants," as rhinestones once were called. One of my favorites was a delicate 1925 art deco comb with a curved handle and its own carrying case; together, they looked like an elegant purse made of tortoise-shell and secured with a rhinestone clasp. Just four inches long, it was surely designed for the short hair of a Jazz Age beauty. Looking at the comb, I could imagine its first owner, a bright spirit in a dropped-waist dress and Louise Brooks bob, reveling in her liberation from corsets, long gowns, and heavy hair buns.

Surprisingly, these gorgeous antiques are quite affordable. Celluloid plastic made it possible, for the first time, to produce combs in real abundance — keeping prices low even for today's collector who doesn't have a lot to spend but wants to own something fabulous. For people at the dawn of the plastic age, celluloid offered what one writer called "a forgery of many of the necessities and luxuries of civilized life," a foretoken of the new material culture's aesthetic and abundance.

Combs are one of our oldest tools, used by humans across cultures and ages for decoration, detangling, and delousing. They derive from the most fundamental human tool of all — the hand. And from the time that humans began using combs instead of their fingers, comb design has scarcely changed, prompting the satirical paper the *Onion* to publish a piece titled "Comb Technology: Why Is It So Far Behind the Razor and Toothbrush Fields?" The Stone Age craftsman who made the oldest known comb — a small four-toothed number carved from animal bone some eight thousand years ago — would have no trouble knowing what to do with the bright blue plastic version sitting on my bathroom counter.

For most of history, combs were made of almost any material humans had at hand, including bone, tortoiseshell, ivory, rubber, iron, tin, gold, silver, lead, reeds, wood, glass, porcelain, papier-mâché. But in the late nineteenth century, that panoply of possibilities began to fall away with the arrival of a totally new kind of material — celluloid, the first man-made plastic. Combs were among the first and most popular objects made of celluloid. And having crossed that material Rubicon, comb makers never went back. Ever since, combs generally have been made of one kind of plastic or another.

The story of the humble comb's makeover is part of the much larger story of how we ourselves have been transformed by plastics. Plastics freed us from the confines of the natural world, from the material constraints and limited supplies that had long bounded human activity. That new elasticity unfixed social boundaries as well. The arrival of these malleable and versatile materials gave producers the ability to create a treasure trove of new products while expanding opportunities for people of modest means to become consumers.

Plastics held out the promise of a new material and cultural democracy. The comb, that most ancient of personal accessories, enabled anyone to keep that promise close.

What is plastic, this substance that has reached so deeply into our lives? The word comes from the Greek verb *plassein*, which means "to mold or shape." Plastics have that capacity to be shaped thanks to their structure, those long, flexing chains of atoms or small molecules bonded in a repeating pattern into one gloriously gigantic molecule. "Have you ever seen a polypropylene molecule?" a plastics enthusiast once asked me. "It's one of the most beautiful things you've ever seen. It's like looking at a cathedral that goes on and on for miles."

In the post–World War II world, where lab-synthesized plastics have virtually defined a way of life, we've come to think of plastics as unnatural, yet nature has been knitting polymers since the beginning of life. Every living organism contains these molecular daisy chains. The cellulose that makes up the cell walls in plants is a polymer. So are the proteins that make up our muscles and our skin and the long spiraling ladders that hold our genetic destiny, DNA. Whether a polymer is natural or synthetic, chances are its backbone is composed of carbon, a strong, stable, glad-handing atom that is ideally suited to forming molecular bonds. Other elements — typically oxygen, nitrogen, and hydrogen — frequently join that carbon spine, and the choice and arrangement of those atoms produces specific varieties of polymers. Bring chlorine into that molecular conga line, and you can get polyvinyl chloride, otherwise known as vinyl; tag on fluorine, and you can wind up with that slick nonstick material Teflon.

Plant cellulose was the raw material for the earliest plastics, and with peak oil looming, it is being looked at again as a base for a new generation of "green" plastics. But most of today's plastics are made of hydrocarbon molecules — packets of carbon and hydrogen — derived from the refining of oil and natural gas. Consider ethylene, a gas released in the processing of both substances. It's a sociable molecule consisting of four hydrogen atoms and two carbon atoms linked in the chemical equivalent of a double handshake. With a little chemical nudging those carbon atoms release one bond, allowing each to reach out and grab the carbon in another ethylene molecule. Repeat the process thousands of times and *voilà!*, you've got a new giant molecule, polyethylene, one of the most common and versatile plastics. Depending on how it's processed, the plastic can be used to wrap a sandwich or tether an astronaut during a walk in deep space.

This *New York Times* dispatch is more than a hundred and fifty years old, and yet it sounds surprisingly modern: elephants, the paper warned in 1867, were in grave danger of being "numbered with extinct species" because of humans' insatiable demand for the ivory in their tusks. Ivory, at the time, was used for all manner of things, from buttonhooks to boxes, piano keys to combs. But one of the biggest uses was for billiard balls. Billiards had come to captivate upper-crust society in the United States as well as in Europe. Every estate, every mansion had a billiards table, and by the mid-1800s, there was growing concern that there would soon be no more elephants left to keep the game tables stocked with balls. The situation was most dire in Ceylon, source of the ivory that made the best billiard balls. There, in the northern part of the island, the *Times* reported, "upon the reward of a few shillings per head being offered by the authorities, 3,500 pachyderms were dispatched in less than three years by the natives." All told, at least one million pounds of ivory were consumed each year, sparking fears of an ivory shortage. "Long before the elephants are no more and the mammoths used up," the *Times* hoped, "an adequate substitute may [be] found."

Ivory wasn't the only item in nature's vast larder that was starting to run low. The hawksbill turtle, that unhappy supplier of the shell used to fashion combs, was becoming scarcer. Even cattle horn, another natural plastic that had been used by American comb makers since before the Revolutionary War, was becoming less available as ranchers stopped dehorning their cattle.

In 1863, so the story goes, a New York billiards supplier ran a newspaper ad offering "a handsome fortune," ten thousand dollars in gold, to anyone who could come up with a suitable alternative for ivory.

John Wesley Hyatt, a young journeyman printer in Upstate New York, read the ad and decided he could do it. Hyatt had no formal training in chemistry, but he did have a knack for invention — at the age of twenty-three, he'd patented a knife sharpener. Setting up in a shack behind his home, he began experimenting with various combinations of solvents and a doughy mixture made of nitric acid and cotton. (That nitric acid–cotton combination, called guncotton, was daunting to work with because it was highly flammable, even explosive. For a while it was used as a substitute for gunpowder until producers of it got tired of having their factories blow up.)

As he worked in his homemade lab, Hyatt was building on decades of invention and innovation that had been spurred not only by the limited quantities of natural materials but also by their physical limitations. The Victorian era was fascinated with natural plastics such as rubber and shellac. As historian Robert Friedel pointed out, they saw in these substances the first hints of ways to transcend the vexing limits of wood and iron and glass. Here were materials that were malleable but also amenable to being hardened into a final manufactured form. In an era already being rapidly transformed by industrialization, that was an alluring combination of qualities — one hearkening to both the solid past and the tantalizingly fluid future. Nineteenth-century patent books are filled with inventions involving combinations of cork, sawdust, rubbers, and gums, even blood and milk protein, all designed to yield materials that had some of the qualities we now ascribe to plastic. These plastic prototypes found their way into a few decorative items, such as daguerreotype cases, but they were really only intimations of things to come. The noun *plastic* had not yet been coined — and wouldn't be until the early twentieth century — but we were already dreaming in plastic.

Hyatt's breakthrough came in 1869. After years of trial and error, Hyatt ran an experiment that yielded a whitish material that had "the consistency of shoe leather" but the capacity to do much more than sole a pair of shoes. This was a malleable substance that could be made as hard as horn. It shrugged off water and oils. It could be molded into a shape or pressed paper-thin and then cut or sawed into usable forms. It was created from a natural polymer — the cellulose in the cotton — but had a versatility none of the known natural plastics possessed. Hyatt's brother Isaiah, a born marketer, dubbed the new material *celluloid,* meaning "like cellulose."

While celluloid would prove a wonderful substitute for ivory, Hyatt apparently never collected the ten-thousand-dollar prize. Perhaps that's because celluloid didn't make very good billiard balls — at least not at first. It lacked the bounce and resilience of ivory, and it was highly volatile. The first balls Hyatt made produced a loud crack, like a shotgun blast, when they knocked into each other. One Colorado saloonkeeper wrote Hyatt that "he didn't mind, but every time the balls collided, every man in the room pulled a gun."

However, it was an ideal material for combs. As Hyatt noted in one of his early patents, celluloid transcended the deficiencies that plagued many traditional comb materials. When it got wet, it didn't get slimy, like wood, or corrode, like metal. It didn't turn brittle, like rubber, or become cracked and discolored, like natural ivory. "Obviously none of the other materials . . . would produce a comb possessing the many excellent qualities and inherent superiorities of a comb made of celluloid," Hyatt wrote in one of his patent applications. And while it was sturdier and steadier than most natural materials, it could, with effort, be made to look like many of them.

Celluloid could be rendered with the rich creamy hues and striations of the finest tusks from Ceylon, a faux material marketed as French Ivory. It could be mottled in browns and ambers to emulate tortoiseshell; traced with veining to look like marble; infused with the bright colors of coral, lapis lazuli, or carnelian to resemble those and other semiprecious stones; or blackened to look like ebony or jet. Celluloid made it possible to produce counterfeits so exact that they deceived "even the eye of the expert," as Hyatt's company boasted in one pamphlet. "As petroleum came to the relief of the whale," the pamphlet stated, so "has celluloid given the elephant, the tortoise, and the coral insect a respite in their native haunts; and it will no longer be necessary to ransack the earth in pursuit of substances which are constantly growing scarcer."

Of course, scarcity has long been key to the collection of qualities that make an object luxurious and valuable. There are few things we long for more insistently than those that are just beyond our reach. The writer O. Henry captured the sting — and ultimate emptiness — of that longing in his 1906 story "The Gift of the Magi." Della, the young wife, falls in love with a set of combs she spies in a store on Broadway: "Beautiful combs, pure tortoise shell, with jewelled rims . . . They were expensive combs, she knew, and her heart had simply craved and yearned over them without the least hope of possession." There is no way Della can afford such combs, not on her husband's twenty-dollar-a-week salary. Nor, it seems, does Della come from a family that might bequeath such exquisite heirlooms. Living in an eight-dollar-a-month flat that looks out on an airshaft, saving pennies "one and two at a time by bulldozing the grocer and the vegetable man and the butcher," Della at the start of the story defines her world by what she lacks rather than what she has. Yet in the end that nagging sense of lack — the driver of modern consumption — is not what motivates Della. On Christmas Eve, she cuts and sells her hair — her proudest possession — to buy a watch fob for her husband's treasured gold watch. Meanwhile, he sells the watch to purchase Della her tortoise-shell heart's desire. In that pair of selfless acts, both define themselves by what they give up — what they don't have — rather than by what they hope to consume.

Had those combs been made of celluloid, O. Henry would have had no story to tell.

Even on husband Jim's modest salary, celluloid combs would have been within reach. Indeed, the irony of O. Henry's tale turns on a notion of generosity that only makes sense in a world of scarce resources and rare commodities. In Plasticville, it's not entirely clear what gifts the Magi might offer. But obviously the possible virtues of scarcity were not on Hyatt's mind when his company enthused that a "few dollars invested in Celluloid" equaled "hundreds expended in the purchase of genuine products of nature."

That fantastic talent for forgery became a hallmark of the celluloid industry. It would have been easier — and less expensive — to omit the painstaking layering and dyeing required to make a comb that appeared to be ivory or tortoiseshell. But custom demanded the appearance of natural materials. People took pleasure in this game of artifice — evidence, in a sense, of humanity's growing mastery over nature. Art critic John Ruskin described the thrill of the trompe l'oeil: "Whenever anything looks like what it is not, the resemblance being so great as *nearly* to deceive, we feel a kind of pleasurable surprise, an agreeable excitement of mind."

Perhaps most agreeable of all was the prospect that one's inexpensive possessions might be regarded as rare by others. Hyatt's company offered an extensive line of toiletry sets promoted with wonderfully ambiguous names such as Ivaleur, Amberleur, Shelleur, and Ebonleur. The company urged its salesmen to emphasize the artistic appeal of these products in hopes of persuading women "who have not already done so as a matter of good taste [to] turn from the ostentatious silver toilet ware to that which is less expensive though really more beautiful."

Thanks to celluloid, anyone — even O. Henry's Della — could now afford to possess a comb, brush, and mirror set that looked as if it might have belonged to a Rockefeller, "with graining so delicate and true," one company boasted, "that you would think it could only come from the gleaming tusks of some fine old elephant." Any shop girl could pin up her hair with gorgeously filigreed forgeries of the carved tortoiseshell combs she could never have afforded. (Good thing too, since, according to one turn-of-the-century observer, contemporary hairstyles often demanded "a couple of pounds of Celluloid" combs.) Material scarcity had stoked the longing of a Della, but celluloid managed to take the pain out of consumer desire — turning the wistful and class-conscious window-gazer into the satisfied shopper. Celluloid helped spread a taste for luxury — or at least the look of luxury — to those who'd never been able to entertain fantasies of the finer life. But even more important, it helped fuel a growing demand for things, period.

Celluloid appeared at a time when the country was changing from an agrarian economy to an industrial one. Where once people had grown and prepared their own food and made their own clothes, increasingly they were eating, drinking, wearing, and using things that came from factories. We were fast on our way to becoming a country of consumers. Celluloid was the first of the new materials that would level the playing field for consumption, as historian Jeffrey Meikle pointed out in his insightful cultural history *American Plastic*. "By replacing materials that were hard to find or expensive to process, celluloid democratized a host of goods for an expanding consumption-oriented middle class." Ample supplies of celluloid allowed manufacturers to keep up with rapidly rising demand while also keeping costs down. Like other plastics that would follow, celluloid offered a means for Americans to buy their way into new stations in life.

Combs obviously weren't the only example of celluloid's democratizing effect. Celluloid collars stamped with the weave of linen allowed any man to look the part of a dandy. Celluloid toothbrushes replaced ones with bone handles, making dental hygiene available for mere pennies. Once Hyatt perfected a way to make celluloid billiard balls, billiards stepped down from the plush cognac-and-cigars milieu and into community halls. No longer just a rich man's pleasure, billiards became an everyman's game, especially when the tables gained pockets and the sport evolved into pool. As *The Music Man*'s Professor Harold Hill sang, "Pockets that mark the diff'rence / between a gentleman and a bum."

Perhaps celluloid's greatest impact was serving as the base for photographic film. The history of film, which is one of plastic's most profound cultural legacies, is a book in itself. Here celluloid's gift for facsimile achieved its ultimate expression, the complete transmutation of reality into illusion, as three-dimensional flesh-and-blood beings were transformed into two-dimensional ghosts shimmering on a screen. Here, too, celluloid had a powerful leveling effect in several ways. Film offered a new kind of entertainment, available to and shared by the masses. A dime bought anyone an afternoon of drama, romance, action, escape. Audiences from Seattle to New York roared at the antics of Buster Keaton and thrilled to hear Al Jolson speak the first words in a talkie: "Wait a minute, wait a minute, you ain't heard nothin' yet." The mass culture of film reeled across class, ethnic, racial, and regional lines, drawing one and all into shared stories and imbuing us with the sense that reality itself is as changeable and ephemeral as the names on the movie marquee. With film, an old elite was dethroned; the glamour once associated with class and social standing was now possible for anyone with good cheekbones, some talent, and a bit of luck. A Della could become a socialite onscreen and a movie star in real life.

Ironically, the world opened by celluloid film nearly killed the celluloid-comb industry. In 1914, Irene Castle, a ballroom dancer turned movie star, decided to cut her long hair into a short bob, prompting female fans across the country to take scissors to their own hair. Nowhere did those shorn locks fall harder than in Leominster, Massachusetts, which had been the country's comb capital since before the Revolutionary War and which was now the cradle of the celluloid industry, much of it devoted to combs. Nearly overnight, half of the comb companies in town were forced to shut down, throwing thousands of comb makers out of work. Sam Foster, owner of Foster Grant, one of the town's leading celluloid-comb companies, told his workers not to worry. "We'll make something else," he assured them. He hit on the idea of making sunglasses, creating an entirely new mass market. "Who's that behind those Foster Grants?" the company later teased in ads that featured photographs of celebrities such as Peter Sellers, Mia Farrow, and Raquel Welch hidden behind dark lenses. With a quick trip to the local drugstore, anyone could acquire the same glamorous mystique.

For all its significance, celluloid had a fairly modest place in the material world of the early twentieth century, limited mainly to novelties and small decorative and utilitarian items, like the comb. Making things from celluloid was a labor-intensive process; combs were molded in small batches and still had to be sawed and polished by hand. And because the material was so volatile, the factories were like tinderboxes. Workers often labored under a constant spray of water, but fires were still common. It wasn't until the development of more cooperative polymers that plastics truly began to transform the look, feel, and quality of our lives. By the 1940s, we had both the plastics and the machines to mass-produce plastic products. Injection-molding machines — now standard equipment in plastics manufacturing — turned raw plastic powders or pellets into a molded, finished product in a one-shot process. A single machine equipped with a mold containing multiple cavities could pop out ten fully formed combs in less than a minute.

DuPont, which bought one of the original celluloid companies in Leominster, released photos in the mid-1930s showing the daily output of a father-and-son pair of comb makers. In the photos, the father is standing next to a tidy stack of three hundred and fifty celluloid combs, while ten thousand injection-molded combs surround the son. And although a single celluloid comb cost one dollar in 1930, by the end of the decade one could buy a machine-molded comb of cellulose acetate for anywhere from a dime to fifty cents. With the rise of mass-production plastics, the fanciful decorative combs and faux ivory dresser sets so popular in the celluloid era gradually disappeared. Combs were now stripped down to the most essential elements — teeth and handle — in service of their most basic function.

Bakelite, the first truly synthetic plastic, a polymer forged entirely in the lab, paved the way for successes like that of DuPont's injection-mold-comb-making son. As with celluloid, Bakelite was invented to replace a scarce natural substance: shellac, a product of the sticky excretions of the female lac beetle. Demand for shellac began shooting up in the early twentieth century because it was an excellent electrical insulator. Yet it took fifteen thousand beetles six months to make enough of the amber-colored resin needed to produce a pound of shellac. To keep up with the rapid expansion of the electrical industry, something new was needed.

As it turned out, the plastic Leo Baekeland invented by combining formaldehyde with phenol, a waste product of coal, and subjecting the mixture to heat and pressure was infinitely more versatile than shellac. Though it could, with effort, be made to mimic natural materials, it didn't have celluloid's knack for imitation. Instead, it had a powerful identity of its own, which helped encourage the development of a distinctively plastic look. Bakelite was a dark-colored, rugged material with a sleek, machinelike beauty, "as stripped down as a Hemingway sentence," in writer Stephen Fenichell's words. Unlike celluloid, Bakelite could be precisely molded and machined into nearly anything, from tubular industrial bushings the size of mustard seeds to full-size coffins. Contemporaries hailed its "protean adaptability" and marveled at how Baekeland had transformed something as foul-smelling and nasty as coal tar—long a discard in the coking process—into this wondrous new substance.

Families gathered around Bakelite radios (to listen to programs sponsored by the Bakelite Corporation), drove Bakelite-accessorized cars, kept in touch with Bakelite phones, washed clothes in machines with Bakelite blades, pressed out wrinkles with Bakelite-encased irons—and, of course, styled their hair with Bakelite combs. "From the time that a man brushes his teeth in the morning with a Bakelite-handled brush until the moment when he removes his last cigarette from a Bakelite holder, extinguishes it in a Bakelite ashtray and falls back upon a Bakelite bed, all that he touches, sees, uses will be made of this material of a thousand purposes," *Time* magazine enthused in 1924 in an issue that sported Baekeland on the cover.

The creation of Bakelite marked a shift in the development of new plastics. From then on, scientists stopped looking for materials that could emulate nature; rather, they sought "to rearrange nature in new and imaginative ways." The 1920s and '30s saw an outpouring of new materials from labs around the world. One was cellulose acetate, a semisynthetic product (plant cellulose was one of its base ingredients) that had the easy adaptability of celluloid but wasn't flammable. Another was polystyrene, a hard, shiny plastic that could take on bright colors, remain crystalline clear, or be puffed up with air to become the foamy polymer DuPont later trademarked as Styrofoam. DuPont also introduced nylon, its answer to the centuries-long search for an artificial silk. When the first nylon stockings were introduced, after a campaign that promoted the material as being as "lustrous as silk" and as "strong as steel," women went wild. Stores sold out of their stock in hours, and in some cities, the scarce supplies led to nylon riots, full-scale brawls among shoppers. Across the ocean, British chemists discovered polyethylene, the strong, moisture-proof polymer that would become the sine qua non of packaging. Eventually, we'd get plastics with features nature had never dreamed of: surfaces to which nothing would stick (Teflon), fabrics that could stop a bullet (Kevlar).

Though fully synthetic like Bakelite, many of these new materials differed in one significant way. Bakelite is a thermoset plastic, meaning that its polymer chains are hooked together through the heat and pressure applied when it is molded. The molecules set the way batter sets in a waffle iron. And once those molecules are linked into a daisy chain, they can't be unlinked. You can break a piece of Bakelite, but you can't melt it down to make it into something else. Thermoset plastics are immutable molecules — the Hulks of the polymer world — which is why you'll still find vintage Bakelite phones, pens, bangles, and even combs that look nearly brand-new.

Polymers such as polystyrene and nylon and polyethylene are thermoplastics; their polymer chains are formed in chemical reactions that take place before the plastic ever gets near a mold. The bonds holding these daisy chains together are looser than those in Bakelite, and as a result these plastics readily respond to heat and cold. They melt at high temperatures (how high depends on the plastic), solidify when cooled, and if made cold enough can even freeze. All of which means that, unlike Bakelite, they can be molded and melted and remolded over and over again. Their shape-shifting versatility is one reason thermoplastics quickly eclipsed the thermosets and today constitute about 90 percent of all the plastics produced.

Many of the new thermoplastics at one time or another found their way into combs, which, thanks to injection molding and other new fabrication technologies, could be made faster and in far greater quantities than ever before — thousands of combs in a single day. This was a small feat in and of itself, but multiplied across all the necessities and luxuries that could then be inexpensively mass-produced, it's understandable why many at the time saw plastics as the harbinger of a new era of abundance. Plastics, so cheaply and easily produced, offered salvation from the haphazard and uneven distribution of natural resources that had made some nations wealthy, left others impoverished, and triggered countless devastating wars. Plastics promised a material utopia, available to all.

At least, that was the hopeful vision of a pair of British chemists writing on the eve of World War II. "Let us try to imagine a dweller in the 'Plastic Age,'" Victor Yarsley and Edward Couzens wrote. "This 'Plastic Man' will come into a world of colour and bright shining surfaces . . . a world in which man, like a magician, makes what he wants for almost every need." They envisioned him growing up and growing old surrounded by unbreakable toys, rounded corners, unscuffable walls, warpless windows, dirt-proof fabrics, and lightweight cars and planes and boats. The indignities of old age would be lessened with plastic glasses and dentures until death carried the plastic man away, at which point he would be buried "hygienically enclosed in a plastic coffin."

That world was delayed in coming. Most of the new plastics discovered in the 1930s were monopolized by the military over the course of World War II. Eager to conserve precious rubber, for instance, in 1941 the U.S. Army put out an order that all combs issued to servicemen be made of plastic instead of hard rubber. So every member of the armed forces, from private to general, in white units and black, got a five-inch black plastic pocket comb in his "hygiene kit." Of course, plastics were also pressed into far more significant service, used for mortar fuses, parachutes, aircraft components, antenna housing, bazooka barrels, enclosures for gun turrets, helmet liners, and countless other applications. Plastics were even essential to the building of the atomic bomb: Manhattan Project scientists relied on Teflon's supreme resistance to corrosion to make containers for the volatile gases they used. Production of plastics leaped during the war, nearly quadrupling from 213 million pounds in 1939 to 818 million pounds in 1945.

Come V-J Day, however, all that production potential had to go somewhere, and plastics exploded into consumer markets. (Indeed, as early as 1943, DuPont had a whole division at work preparing prototypes of housewares that could be made of the plastics then commandeered for the war.) Just months after the war's end, thousands of people lined up to get into the first National Plastics Exposition in New York, a showcase of the new products made possible by the plastics that had proven themselves in the war. For a public weary of two decades of scarcity, the show offered an exciting and glittering preview of the promise of polymers. There were window screens in every color of the rainbow that would never need to be painted. Suitcases light enough to lift with a finger, but strong enough to carry a load of bricks. Clothing that could be wiped clean with a damp cloth. Fishing line as strong as steel. Clear packaging materials that would allow a shopper to see if the food inside was fresh. Flowers that looked like they'd been carved from glass. An artificial hand that looked and moved like the real thing. Here was the era of plenty that the hopeful British chemists had envisioned. "Nothing can stop plastics," the chairman of the exposition crowed.

All those ex-GIs with their standard-issue combs were coming home to a world of not only material abundance but also rich opportunities created by the GI Bill, housing subsidies, favorable demographics, and an economic boom that left Americans with an unprecedented level of disposable income. Plastics production expanded explosively after the war, with a growth curve that was steeper than even the fast-rising GNP's. Thanks to plastics, newly flush Americans had a never-ending smorgasbord of affordable goods to choose from. The flow of new products and applications was so constant it was soon the norm. Tupperware had surely always existed, alongside Formica counters, Naugahyde chairs, red acrylic taillights, Saran wrap, vinyl siding, squeeze bottles, push buttons, Barbie dolls, Lycra bras, Wiffle balls, sneakers, sippy cups, and countless more things. The nascent industry partnered with the press, especially women's magazines, to sell consumers on the virtue of plastics. "Plastics are here to free you from drudgery," *House Beautiful* promised housewives in a special fifty-page issue in October of 1947 titled "Plastics . . . A Way to a

Better, More Carefree Life." Even combs were brought into the service of consumption, taking on a new function as mini-billboards for various companies. Hotels, airlines, railroads, and other industries in the late 1950s began handing out complimentary combs stamped with the companies' names.

That proliferation of goods helped engender the rapid social mobility that took place after the war. We were a nation of consumers now, a society increasingly democratized by our shared ability to enjoy the conveniences and comforts of modern life. Not just a chicken in every pot, but a TV and stereo in every living room, a car in every driveway. Through the plastics industry, we had an ever-growing ability to synthesize what we wanted or needed, which made reality itself seem infinitely more open to possibility, profoundly more malleable, as historian Meikle observed. Now full-fledged residents of Plasticville, we began to believe that we too were plastic. As *House Beautiful* assured readers in 1953: "You will have a greater chance to be yourself than any people in the history of civilization."

A Throne for the Common Man

I N 1968, NEW YORK'S Museum of Contemporary Crafts put on a landmark exhibit showcasing art, furniture, housewares, jewelry, and sundry other items made of plastic. The show, "Plastic as Plastic," was meant as a tribute to the new kinds of artistic freedom made possible by polymers. As the *New York Times* art critic Hilton Kramer wrote in his review of the show, here "was the answer to an artist's dream" — an "entire family of materials that can be made to assume virtually any size, shape, form, or color the mind of man may conceive." Was it any wonder that artists and designers had fallen deeply in love with these new materials?

And yet, Kramer was struck by how lamely the artists in the show had responded to that "almost Faustian freedom," at least in comparison with the industrial designers, those creative minds responsible for translating aesthetic visions into real-world applications. In his view, the designers, especially the ones dreaming up furniture, were "so evidently more relaxed, more inventive, and more inspired in the world of plastics than even the finest of the artists." Their creations were the ones that were "defining a new world of feeling for us."

Designers had been exploring that new world for decades by the time of the exhibit. Since the arrival of Bakelite, they had seen in plastics the opportunity to invent a modern aesthetic for everyday life, whether in cars, coffeepots, or chairs. In fact, especially chairs. If the comb brought plastic to the masses, the chair showed us how fabulous plastic could be.

Until recently, I never gave much thought to chairs, other than assigning vague comfort ratings to the various ones in my life. But as I've come to appreciate, it takes a lot of ingenuity to make a good chair. There's a reason the Herman Miller company reportedly spent ten million dollars developing its ergonomically exquisite Aeron office chair.

We're more intimate with chairs than with nearly any other piece of furniture. Yet the same dining-room chair in which I plant the ample bottom of my five-foot-three-inch frame also has to be able to accommodate my skinny, nearly buttless six-foot husband, my fast-growing teenage sons, and my petite preteen daughter. A chair has to support all shapes and sizes and still be reasonably comfortable. That's a tall order to fill. No other piece of furniture has so many demands placed upon it.

As a result, the chair has long been considered the Mount Everest of furniture design. Time and again, creative minds have tackled this seemingly simple item, looking for new and innovative ways to marry form and function. Design museums are filled with chairs, as are design-history books. "Both from a design standpoint and an anthropological standpoint, chairs are extremely important," said Paola Antonelli, curator of the design department at the Museum of Modern Art.

If you look back at the history of chairs, it's remarkable how consistent the fundamental form has been. The oldest known chair — a 3,400-year-old specimen unearthed from the tomb of the Egyptian queen Hetepheres (and intended to provide a good seat for whatever events awaited her in the afterlife) — would be more or less at home in a modern-day living room. It's wide and low, with high arms and four legs ending in feet carved to look like lion's paws. That basic rectilinear form recurs through the centuries, across countries and cultures. It's a shape dictated by both the features of the seated human body and the constraints of the materials at hand, which for most of human history were wood and metal, leather and rope, and, only fairly recently, fabric.

Even with that limited vocabulary, chairs offer eloquent testimony to a culture's Zeitgeist. Consider the chairs of two very different eighteenth-century worlds. A Louis XIV armchair — gilded, ornate, richly detailed — mirrors the pomp and politics of the Versailles court in the same way the sober, plain lines of a Shaker chair portray that sect's austere faith. The baroquely decorated armchair reflected the Sun King's glory — and only he was allowed to enjoy it; all other members of the court were forced to sit on footstools. The Shaker craftsmen deliberately avoided adornment; the only detailing was that which served a practical purpose. Their attitude toward their furnishings, as historians of Shaker architecture have noted, "was no more sentimental than their attitude toward their own flesh-and-blood bodies. It was the spirit of usefulness within that mattered, not the vessel itself."

We can see the flowering of an expansive, creative Hellenistic culture in the graceful, curving lines of the fifth-century klismos chair, and the heavy hand of feudal rule in the massive, blocky thronelike chairs of medieval Europe. The spirit of industrialization is apparent in the brilliant design of what is now the classic wooden café chair. The Thonet Model 14, as aficionados know it, was introduced in 1859 by German cabinetmaker Michael Thonet, who was determined to create a chair that could be mass-produced and sold for an affordable price. He succeeded by reducing the geometry of a good chair to half a dozen easily assembled parts: two wood circles, two sticks, a pair of bentwood arches, plus ten screws and two nuts. By 1930, fifty million had been sold, and millions more have been sold since. The chairs of today are equally telling. Americans' obsession with ergonomic seating fairly shouts that we are an exceedingly stationary people.

But chairs are not only cultural artifacts; they have long served as artistic canvases. As industrial designer George Nelson once observed, "Every truly original idea — every innovation in design, every new application of materials, every technical innovation for furniture — seems to find its most important expression in a chair." From the mid-twentieth century on, much of that innovation was inspired by plastics. The arrival of this chemical armada blew away many of the constraints imposed by traditional materials. Chair designers could develop ways to conform a seat to the human body beyond just a series of right angles. A chair of plastic could have many legs or, like a beanbag chair, none. It could be hard or squishy or filled with air; it could be shaped like a baseball or a baseball glove. Polymer technology permitting, the only limits were a designer's imagination.

The Greek root of the word *plastics* can be used as an adjective or a verb but not as a noun, which is probably truer to the nature of plastic than anyone imagined when the word was first coined. For although we talk about plastic as a thing, it doesn't have the thingness, the kind of grounded organic identity, found in natural substances. Wood, stone, metal, mineral: all have innate properties that dictate how we use them and how we think about them. We know that a diamond will be hard enough to scratch glass, that a gold ring won't rust, and that a piece of ebony can be polished to a high sheen. And when we look at an object made of natural materials, we see hints of how it came into being, whether it was planed or hammered or forged or woven.

But a piece of plastic is essentially inscrutable, offering few clues as to its past or future. Though specific properties may be engineered into any given polymer, the only innate quality defining plastics as a family of materials is . . . their plasticity, their protean ability to be whatever we need them to be. As the French philosopher Roland Barthes observed in a famous 1957 meditation on plastic: "the quick-change artistry of plastic is absolute: it can become buckets as well as jewels."

The arrival of such accommodating substances gave us, to the greatest extent ever, the means to shape the world to our wills and whims, our needs and dreams. The makers of Bakelite sought to drive that point home in choosing as their trademark the mathematical symbol for infinity.

Designers were enthralled by that universe of possibility from plastics' earliest days. They loved the design freedom that synthetics offered and the spirit of modernity the materials embodied — a doors-wide-open sensibility that one German critic called *Plastikoptimismus*. To furniture designer Paul T. Frankl, a material like Bakelite spoke "in the vernacular of the twentieth century . . . the language of invention, of synthesis," and he urged his fellow designers to use their full imaginative powers to explore the new materials' frank artificiality. As interpreted by Frankl and other designers working with Bakelite in the '30s and '40s, that was the language of streamlining, a lingo of curves and dashes and teardrop shapes that created a feeling of speed and motion in everyday objects from telephones to radios to martini shakers. Streamline a fountain pen and even that stolid item declared: we're hurtling toward the future here! The infinitely moldable thermoplastics that later became available offered an even broader design vocabulary.

There was another reason designers embraced plastics. From the mid-twentieth century on, modern design has been guided by an egalitarian gospel, a belief that good design needn't cost a lot of money, that even the most mundane items could be things of beauty. "Get the most of the best to the most for the least" was the way Ray and Charles Eames put it in their famous tongue-twisting credo. Plastics were the ideal medium for that mission: malleable, relatively inexpensive, and made for mass manufacture. Or as Karim Rashid, a contemporary designer renowned for his love of synthetic materials, expressed it more recently, "To create a beautiful democratic design, plastic is the best material." The Museum of Modern Art in 1956 acknowledged plastic's contribution to the mission when it included a number of pieces of Tupperware in an exhibit of outstanding twentieth-century design. According to Alison Clarke, a historian of Tupperware, the pieces were praised for being well made and well proportioned, for their "uncluttered" and "carefully considered shapes [that were] marvelously free of the vulgarity that characterized so much household equipment." Just as plastics had helped democratize consumer goods, they now aided in democratizing design.

Yet, as in any new relationship, there were risks. It was all too easy to exploit plastics' powers of mimicry to produce the kinds of gauche imitations — pseudo-wood cabinets and faux-leather recliners — that contributed to the growing reputation of plastic as an inferior material. Plastics' adaptability and glibness undermined their capacity to achieve "dignity" as legitimate materials worthy of being taken seriously, one critic wrote.

This impression was exacerbated by people's unfortunate experiences with plastics in the immediate postwar years. The nascent industry had promised "test-tube marvels," but the peacetime markets were glutted with chemical mishaps. Manufacturers were still on the steep part of the learning curve, and that did not always make for happy consumers. There were plastic plates that melted in hot water, plastic toys that cracked on Christmas morning, plastic raincoats that grew clammy and fell apart in the rain. Polymer technology improved during the 1950s as manufacturers figured out how to make better plastics and, even more important, how to match the right polymer with the right application. But the damage to plastic's reputation had been done.

The designer Charles Eames was well aware of the challenges posed by getting involved with plastics. He and his wife, Ray, had created one of the first iconic plastic chairs — the famed bucket chair — which was made from a curl of fiberglass perched atop a crisscross of thin metal legs. In a lecture to students in 1963, he talked about the differences between working with a natural substance such as granite and a synthetic material such as fiberglass. Granite, he said, is such a hard material that while it might not be easy to create something good out of it, "it is extremely difficult to do something bad.

"Plastic," he continued, "is a different matter. In this spineless material it is extraordinarily easy to do something bad — one can do any imaginable variety of bad without half trying. The material itself puts up no resistance, and whatever discipline there is, the artist must be strong enough to provide." Eames said he viewed plastic in much the same way the Aztecs viewed hard liquor — a means of self-expression too dangerously intoxicating for the young. Under Aztec law, only age and maturity earned someone the right to indulge; young adults who got drunk could be punished with death. Likewise, Eames believed that "plasticene and the airbrush should be reserved for artists over fifty."

The Danish designer Verner Panton was barely thirty when he began dreaming of plastic chairs, in the mid-1950s. He was a recent graduate of architecture school: an ambitious iconoclast with a wild imagination who refused to compromise his beliefs. During the war, he not only opposed the Nazi occupation of Denmark, he quit college to join the Danish resistance, and he spent months in hiding after weapons were found in his apartment. After the war, he moved to Copenhagen to study architecture. He soon found his way into the city's influential design scene, making friends with several of its luminaries.

Yet he had little interest in the quiet, low-key Danish modern look that was then filling middle-class living rooms around the world. He dreaded what he called "grey-beige conformity," and he thought the color white so boring it "should be taxed." He dressed only in blue. He was uninspired by wood and natural fibers. His mind's eye saw space-age shapes, garish hues from the far reaches of the color wheel, twisty, bendy forms that couldn't be achieved through traditional woodworking. Like many of his contemporaries, he was fascinated by the raft of new materials — steel wires, molded plywood, and, most important, plastics — surfacing in the wake of the war. Designers "should now use these materials to create objects which up to now they could only see in their dreams," he told an interviewer. "Personally I'd like to design chairs which exhaust all the technical possibilities of the present."

In fact, he already had one in mind — a radical, very unchairlike chair design he hoped to execute in plastic. But he knew he wouldn't find an audience for his vision in conservative Copenhagen, where, as one prominent designer said, "we have not bothered about anything but changing the kind of wood."

In the late 1950s, Panton bought a Volkswagen van, outfitted it as a mobile studio, and began taking periodic trips across Europe to drop in on designers and manufacturers and distributors he hoped might buy his designs. By the early '60s, he was gaining a reputation for playful imaginative designs that drew on nontraditional materials: he furnished a hotel lobby in the first inflatable plastic furniture; designed UFO-shaped lamps and walls of bubbly backlit plastic panels; created chairs from cylinders of sheet metal. He was also gaining a reputation as an enfant terrible who liked to provoke his more conventional colleagues. At a design fair in 1959 he insisted on attaching the furniture he was exhibiting to the ceiling. He thought it would give visitors a better view of the display. But his fellow exhibitors weren't amused by such attention-grabbing antics. "Many of the artists refused to speak to me for some time after," he recalled.

During each trek across Europe in his VW, Panton brought along a pintsize model of the radical chair he'd envisioned years earlier, hoping to meet someone who was willing to underwrite its production. But for years he couldn't find a willing partner. "It is at most a sculpture, but not a chair," sniffed one furniture maker Panton approached.

It certainly didn't resemble any conventional chair. There were no arms or legs, just a long S-curve of plastic shaped like the silhouette of someone sitting. The seat was supported by a concave vault beneath it. Panton didn't invent that form; it had been pioneered decades earlier by Dutch architect Mart Stam and then popularized by Marcel Breuer, who made a cantilevered chair in chrome tubing and wood. But his vision brought the form into the synthetic age. It would be all sinuous line and shiny surface, a double twist of body-hugging curves that could only be achieved in plastic.

But it wasn't just the shape that kept Panton doggedly making the rounds with his model. He was also captivated by the challenge of creating a chair that could be mass-produced in one step from a single piece of plastic. He got the idea from watching a plastic bucket being made by an injection-molding machine. Plastic pellets went in one end and were quickly melted into a liquid that was shot into a mold, where it was shaped and cooled. Panton was impressed by the speed of the process and by the bucket's low price. If he could make a chair in the same way, he'd accomplish a goal that designers had been chasing for generations: a chair that was literally all of a piece. Such a chair would be the perfect embodiment of the modern industrial age: a harmonious ensemble of form, material, and manufacturing technique, what designers call total design unity.

Total design unity is the ultimate ambition of the design world. It's valued on aesthetic grounds and also because it represents the most efficient way of manufacturing an object. "If you're thinking about how to get good design to the masses in a way that's affordable, single-material forms make the most sense," explained furniture historian Peter Fiell. Panton wasn't the only one caught up in that challenge. Mid-twentieth-century designers across Europe and North America were exploring ways to mass-produce plastic chairs with the new polymers available, such as fiberglass and polypropylene.

Yet, to the disappointment of Panton, technology lagged behind artistic vision. In 1957, his colleague Eero Saarinen dreamed up his famous Tulip chair. He imagined a gently curving petal of white fiberglass unfurling from a slender pedestal. (His goal was to get rid of the "slum of legs" afflicting traditional furniture.) Saarinen wanted seat and pedestal molded *"all of one thing,"* as he later wrote. "All the great furniture of the past, from Tutankhamen's chair to Thomas Chippendale's, have always been a structural whole." But there was a problem: a thin stem made of fiberglass wasn't strong enough to support a sitter's weight. So Saarinen had to settle for making the base out of metal and coating it in white plastic. The chair had the look Saarinen wanted, but he was still disappointed. He told colleagues he would keep looking forward to the day "when the plastic industry has advanced to the point where the chair will be one material, as designed."

If there were hurdles to making an all-plastic chair, designers and manufacturers, especially in Europe, were discovering ways to apply their avant-garde visions to less challenging everyday items, taking advantage of advances in both polymer engineering and plastics processing.

No one was better at this than the Italian company Kartell, maker of the first plastic bucket, arguably the most important application of plastic ever. (When you consider how many eons humans have sought a reliable way to contain and carry water, it's no surprise that buckets are among the first plastic objects to be embraced by traditional societies.) The company was founded in 1949 by chemical engineer Giulio Castelli and his wife, Anna, an architect. They understood the need to improve plastics technology. They were continually searching for new ways to tweak polymers' properties and worked closely with machinists and molders to improve molding processes.

They started out making auto accessories but soon gravitated toward more artistic endeavors. The Castellis recognized early on that plastic materials, unlike natural ones, "acquire an identity . . . only by means of the project itself." Success hinged on the design. So they recruited topflight designers for even the most mundane objects. In Kartell's hands, flyswatters, juicers, ashtrays, lamps, and storage containers acquired an elegant beauty. A standing dustpan designed by Gino Colombini had such geometry and grace that it wound up in a number of design-museum collections.

The Castellis' genius was to take plastic at face value. Unlike so many American manufacturers, they didn't try to deploy it as a substitute for a natural material. They didn't rake it with a wood-like grain, stipple it with the pebbly texture of leather, or sprinkle it with glitter to give it the glow of gold. They let plastic be plastic. The products emerging from their Milan factory boasted bright primary colors, sleek surfaces, crisp Euclidian shapes, undulating curves. It was a style so unabashedly artificial that, as Meikle wrote, "the odor of the refinery seemed to linger" on each item. Not everyone appreciated the look, but it was indisputably a style, one fully grounded in the slippery nature of the material. Kartell's designs made it possible for people to believe that plastic, like traditional materials, had some noble essence.

But even Kartell had trouble creating a one-piece chair. For the factory to make a full-size chair, the molds had to be massive, the machinery needed to house and press them together even more so. Some designers came close but were always stymied by the problem of those cursed four legs. Marco Zanuso and Richard Sapper dreamed up a child's chair made of polyethylene for Kartell. The company could mold the back and seat all in one piece, but the legs had to be produced and attached separately. Joe Colombo hit the same wall when he designed an adult-sized chair for the company in 1967.

Panton's legless chair, however, posed fewer production challenges. It was a better fit for plastic — or at least for the state of plastics processing at the time.

The exact history of his chair is not well documented; Panton himself gave contradictory accounts of how it finally came about. What is known is that in the late '60s he finally found a partner willing to take up production of the S chair — a Swiss company that made Herman Miller furniture under license. The company's owner wasn't wild about Panton's design, but his son, Rolf Fehlbaum, was. "It's interesting, it's new, it's exciting," Fehlbaum told his father, urging him to take it on.

The chair proved more challenging than Panton or his new partners had expected. For a few years they experimented with materials and processes, working closely with plastics manufacturers, who were eager to participate in what they all recognized was a groundbreaking project. In 1968, they found the perfect plastic for their project: a new, glossy hard polyurethane foam made by Bayer and called Baydur. Later that year, the company began producing the seat that would go down in design history.

Sleek, sexy, and a technical first, the Panton chair, as it came to be called, was an instant success — at least in the world of design. To Panton's enduring disappointment, the chair was never a huge commercial success; it was a little too weird for the average middle-class consumer with a living room furnished in American Colonial. Nonetheless, it quickly gained status as the iconic chair of the era, the embodiment of sixties exuberance and openness to experimentation. To Mathias Remmele, who curated a museum exhibit of Panton's work, the chair captured something even deeper: "It embodies the enthusiasm of an era in which society's faith in progress and in the supremacy of technology over matter was still largely unshaken." In this incarnation, plastic was cool. The chair graced the cover of design magazines and was recruited for ads where it could lend its sex appeal to unsexy products like dishwashers. One magazine featured a model posing provocatively with a glossy red Panton in a photo spread entitled "How to Undress in Front of Your Husband."

In the wake of the Panton chair, designers came up with even trippier concepts: Inflatable living-room sets. Seats shaped like huge molars, oversize bananas, lips, sea urchins, even a giant patch of grass. One day somewhere around 1970, my solidly midwestern mother came home with a shiny brown vinyl ottoman in the shape of a mushroom. The Panton has gone in and out of fashion. Now it's in again, rejuvenated by the mid-twentieth-century-focused furniture retailer Design Within Reach, which mass-produces the chair in great numbers using a less costly plastic, polypropylene.

Whatever the chair's status as a pop-art icon, the most important thing about it is the simple fact of its creation. As furniture historian Peter Fiell said emphatically, when that first chair fell from that massive mechanical womb, fully formed but untouched by human hands, it was "the single most important moment in the history of furniture since the dawn of civilization." (It's the sort of sweeping judgment one is allowed to make when one has written a book called *1000 Chairs*.) Panton and his partners had figured out the difficult union of form and material and manufacture. They had achieved total design unity. Or, as Fiell put it, "They'd found the holy grail."

The temptations of plastic being what they are, it was only a matter of time before that holy grail would devolve into a Dixie Cup. For, technologically speaking, it's more or less a straight line from the highbrow Panton chair to the lowbrow plastic chair that you can buy today at your local hardware store.

Plain, lightweight, and usually white or green, the monobloc chair (so called because it is molded from a single piece of plastic) may well be the most successful piece of furniture ever invented. Huge flocks of the chairs appear without fail every spring. A basic model costs about the same as a six-pack of Bud.

There are hundreds of millions of the chairs out there, populating the world's porches, poolsides, and parks. They may not show up in design spreads, but as students of the monobloc have observed, look closely, and you're bound to spot them in news stories and photos. Kenyans rose from monoblocs to applaud when Obama's election was announced. There were monoblocs peeking out from Saddam Hussein's hidey-hole, from the prisoners' hell at Abu Ghraib, and in the horrific video of the Baghdad decapitation of American contractor Nicholas Berg (which at least one conspiracy-minded blogger took as evidence proving that the United States was somehow involved in Berg's killing).

White plastic chairs floated up in the debris of both Hurricane Katrina and the Indonesian tsunami. Photos show them at rallies in Cuba, riots in Nigeria, and Chinese celebrations of sixty years of Communist rule. They're in cafés in Israel and in the coffeehouses of its surrounding antagonists Jordan, Syria, and Lebanon. They've been spotted in reclusive North Korea, where even that icon of global commerce Coca-Cola is banned.

The chairs won the world's hearts — and bottoms — because they are inexpensive, light, washable, stackable, and maintenance-free. They can weather any weather. If you don't feel like hosing the schmutz off last year's model, it's easily replaced. They're also reasonably comfortable.

Though the monobloc is descended from the Panton chair, its precise lineage is uncertain. Depending on whom you talk to, the chairs first appeared in the early or late 1970s or the early 1980s, in France or Canada or Australia. Even if the origins of the first monobloc remain obscure, it's not hard to imagine how the breed came into being. Somewhere far beyond the rarefied realm of design, probably in Europe, a utilitarian-minded businessman realized it would be possible to mass-produce plastic chairs. He (this isn't a business with many *shes*) would employ the same injection-molding process that Panton had pioneered. But instead of using an expensive high-tech polymer as Panton had, he would deploy one of the lower-priced commodity plastics, like polypropylene. By this time, the patent on the polymer had expired, and the raw plastic could be had for less than twenty cents a pound. Instead of using an avant-garde design like the Panton chair's, he would revert to a conventional four-legged form, which manufacturers like Kartell finally mastered following Panton's breakthrough. And rather than produce just a few thousand chairs at a time, he would make hundreds of thousands, even millions, which would allow him to recoup the large initial capital costs. Though monobloc chairs are cheap, the equipment to make them is not. An injection-molding press can cost $1 million, while the cost of a new mold can run $250,000 or more.

Indeed, this is the strategy, more or less, that was followed by the French company Allibert in 1978 when it introduced the Dangari, a single-piece plastic garden chair designed by Pierre Paulin, one of France's top furniture designers. The chair was a bestseller. It was more elegant and weighty than today's monoblocs, and it sold for a much heftier price. But at least superficially, it may have served as a model for the lightweight, less thoughtfully designed chairs that soon began flooding the world's markets.

After seeing plastic chairs at a trade show in the early 1980s, Canadian businessman Stephen Greenberg became one of the first North Americans to jump into the monobloc business. It was clear to him that the chairs offered many advantages over the metal garden furniture he was then selling. Plastic chairs wouldn't rust. They stacked easily. The design was brilliantly functional. He began importing monoblocs from France. At the time, he said, there were only a handful of companies on the scene, mostly in Europe. But over the course of the 1980s, that changed, especially after cheaper, used chair-making molds became available. Instead of having to shell out hundreds of thousands of dollars to get in on the monobloc boom, a processor could get himself set up for maybe fifty thousand dollars. Suddenly it seemed like every yahoo with an injection-molding press was producing chairs. Local manufacturers began popping up all over the world—in Argentina, Indonesia, Mexico, Thailand, Israel, New Zealand. Greenberg quit importing monoblocs and began manufacturing them himself. "At our height we were selling five million chairs a year. And we were just one of many. We knew guys in Italy who were producing fifty thousand a day," he told me.

With that kind of volume, the field became viciously cutthroat. Producers kept ratcheting down the price, creating impossibly thin profit margins. While the earliest monoblocs sold for fifty or sixty dollars, by the mid-1990s, they cost a tenth of that. "Eventually a lot of people just put themselves out of business," recalled Greenberg. It was a "sort of suicide." The same story played out in the United States, where intense competition eventually winnowed the number of manufacturers down from the dozen or so in the mid-1980s to the three still making monobloc chairs today.

If you walk into your local hardware store and buy a plastic chair, chances are it was made by Grosfillex, a French veteran of the plastic-furniture business that has a factory in Pennsylvania; U.S. Leisure, the American subsidiary of a huge Israeli plastics conglomerate; or Adams Manufacturing, a privately held company in Portersville, Pennsylvania, a tiny town north of Pittsburgh, with a population of 268. Bill Adams, the founder of Adams Manufacturing, was a relative latecomer to the plastic-chair business, diving into it in the late 1990s. Because of the brutal economics of plastic chairs, his family considered the decision so financially risky that his wife eventually divorced him over it, and his son left the company for several years. Still, Adams had no regrets. As far as he was concerned, there was nothing better he could do for the world than make plastic furniture.

As you'd expect, most plastics manufacturers are gung ho about their products. But for pure and uncomplicated devotion to Team

Polymer, few of them could match Adams, as I discovered when I visited him. "Plastic is so much better than anything else!" he exclaimed in a typical riff. "You can do so much with it. It's so efficient. And it's so clean." His deep and dedicated plastiphilia evoked the unalloyed enthusiasm of the mid-twentieth century. No amount of bad press could shake his conviction that "plastic is just a good thing." When I mentioned growing public concern over litter from plastic bags, he asked skeptically, "Did you see any plastic bags on your way up here?" In all his years vacationing on the Maryland shore, he said, he never once saw plastic trash on the beach, so he didn't believe plastic debris in the ocean posed a problem. Like plastic itself, his faith in polymers was not easily broken down.

Adams was tall and balding with the heavy-lidded, avuncular look of the actor Bert Lahr (the lion in *The Wizard of Oz*). In his sixties when I interviewed him, Adams remembered precisely when he fell for plastic: he was twelve years old, and someone gave him one of those little change purses that you have to flex to open. It was made of vinyl. "I said, 'That's the most amazing thing I've ever seen. This is absolutely beautiful stuff.'"

Still, it was a long, circuitous road from early crush to true commitment. In the late 1970s, Adams was working as a children's librarian but itching to do something different. A born entrepreneur and inventor, he had come up with "this gizmo" that he thought could solve soaring heating bills: bubble wrap attached by thumbtacks to suction cups, a contraption that could seal windows and prevent heat from escaping. Using retirement savings and a modest inheritance, Adams began trying to peddle the gizmo. He met little success until one day he passed by a gas station where duct tape had been used to hang a bunch of signs in the windows. *It's going to take a lot of work to scrape all that tape off,* he thought to himself. *If they just had my suction cups* . . . He stepped inside and had barely begun his spiel before the store manager stopped him and said he'd take two boxes. The next day he visited more gas stations and came home with his wallet stuffed full of dollar bills. Soon, he was taking his suction cup–thumbtack combo to hardware stores all across the Mid-Atlantic region. "People were using them for everything, every time they had something to hang up," he recalled. The proverbial light bulb went on: "I realized no one in the world was taking suction cups seriously. So I started taking suction cups seriously." He bought new equipment, learned how to make suction cups faster and better, and branched out into new suction-cupping opportunities, such as systems for hanging Christmas wreaths and lights. Before long, he had filed more than a hundred and fifty suction-cup-related patents; he'd become America's suction-cup king.

After several years, Adams began to worry that his suction-cup business was too seasonal; he wanted to diversify into products that would keep his factory busy year-round. He heard about a guy going out of business who was selling molds to make folding plastic tables. Adams decided to buy them. Later, he expanded the line with folding chairs and stools. He landed major sales contracts with Walmart, Kmart, and hardware stores. Soon he was wearing a new crown: the world's largest maker of folding plastic furniture. Then he realized there was an even larger empire to conquer: monobloc plastic chairs.

Telling his story, Adams made it sound as if he stumbled from one fortuitous situation to another. Yet given the unforgiving economics of the plastic-chair business, he was obviously a very shrewd businessman. For in just a few years, Adams rose to become one of the country's top producers of monobloc chairs, supplying big-box stores and hardware chains across much of the eastern United States. By the time I met him, in 2008 — four years into the business — he was turning out close to three million chairs a year.

Touring Adams's factory and cavernous warehouses with him, I could see that his pride in his product was not just a nine-to-five show. He truly saw the plastic chair as a thing of beauty, a marvel of utility. Indeed, he had furnished his kitchen and dining room with the plastic chairs and tables that he made. He chose his sage green Mission model, which resembled Mission-style furniture only insofar as the chair had straight back spokes. "I just love plastic furniture," he said earnestly. "There's an elegance to it. If you go back in the history of furniture, to its very beginning, there is nothing that combines chemistry and physics and mechanics and design and style the way [plastic] furniture does."

Adams was not alone in his admiration of the monobloc. In 2001, Jens Thiel, a German management consultant and design buff, started a website devoted to the chair. It has registered as many as thirty thousand hits a month. (It also links to several photo-sharing websites where enthusiasts post pictures of monobloc chairs from around the world.) Thiel got interested in the monobloc when he noticed people were sitting in them at a high-end art show and was struck by the incongruity. Thiel didn't try to defend the chair aesthetically, but he appreciated its simple functionality: "I like them. I find them very practical. I have six monoblocs at my dining-room table."

While the industry has become concentrated in the hands of large corporations in the United States and Europe, elsewhere in the world monoblocs are the products of local enterprises. Around the globe there are an estimated one hundred manufacturers turning out at least five hundred variations on the basic form. I use the word *variations* loosely. There are differences in color — Asian and Latin American countries love bright, vividly colored chairs — and in superficial decorations. Still, producers the world over rarely stray far from the essential design. Given the vast possibilities presented by plastics, I wondered why.

"Ultimately it comes down to price" was the succinct explanation offered by George Lemieux, an Indiana-based consultant who spent more than twenty-five years in the plastics industry, much of it in plastic furniture. The design of monoblocs is largely the result of a series of price calculations driven by consumer demand: how to achieve "the most safe and stable geometry" with the minimum amount of material. There have to be several slats in the back to ensure the chair doesn't buckle when someone leans against it. The legs are splayed at precisely determined angles to prevent them from collapsing outward or folding inward. Corners are curved because that adds a measure of strength. The seat is at least three-sixteenths of an inch thick because that is the minimum thickness needed to support

225 pounds, the benchmark weight of industry standards. In short, the chair is precisely engineered to deliver the safest stable seat for the lowest price — and nothing more.

The economics of monoblocs are so tight, it's difficult to modify the design with anything more than superficial flourishes. For instance, a model that Adams custom-produced for Kmart had a panel of embossed roses across the back. It was profoundly ugly — even cheaper-looking than the basic monobloc, perhaps because the roses truly had nothing to do with the chair's overall design.

Years ago, when Lemieux worked for the chair maker U.S. Leisure, he hired a designer to bring a new look to some of the company's chairs. The innovations seem absurdly minor, but the way they were received is telling. For instance, they tried to introduce a "Southwestern" chair. "It had some unique designs," Lemieux remembered, "little stars and half-moon shapes and different things like that in the back which gave it a southwesterny look. And then we put some flecking in [the plastic] so that it would sparkle like sandstone. It was a nice look."

But one that was quickly crushed in the intensely competitive plastic-furniture market. The chairs sold for $9.99, which according to Lemieux was about two dollars more than consumers were willing to pay for a plastic chair. His company quickly retired the model.

Plastic schlock is not, of course, what Panton and Saarinen and other pioneers of plastic design had in mind when they set out to create a mass-produced plastic chair, a throne for the common man. Yet schlock becomes virtually inevitable when the ethic animating the higher goal is stripped away. Take out the design ethic — not just the aesthetic sense, but the sense of purpose — and all that's left is mass-manufacturability. The result is chair as simple commodity. Useful, affordable, but as soulless as a traffic cone.

Yes, a plastic chair could be anything its producer wanted it to be. But for it to survive the demands of a modern market, the thing it most needed to be was cheap. Manufacturers delivered cheap plastic chairs, therefore consumers expected cheap plastic chairs, therefore manufacturers delivered cheap plastic chairs. It's a pattern — some might say a vicious circle — that makes the plastics-design revolution look more like a commercial putsch. Today's flood of cheap, disposable products mocks early utopian hopes that plastics would fulfill all our wants and needs. Instead of feeling fulfilled, we now often feel choked by an empty abundance.

Today, monoblocs are seen as irredeemably tacky, the emblem of a Walmartized world. A design-minded friend of mine who was planning a party had a nightmare that her husband had filled their house with white monobloc chairs. She woke up in a sweat. *Washington Post* writer Hank Stuever summed up the scorn of many when he wrote: "The resin stacking chair is the Tupperware container of a lard-rumped universe."

I asked various design experts why monoblocs are so widely reviled, and their answers bordered on the metaphysical. "It's almost as if one can feel the cheap thought in the product," said the Panton chair maker Rolf Fehlbaum, now CEO of Vitra Design. It suggests "a moral minimum: how can you make it as cheap as possible so it lasts a few years and then you throw it away." You know, he added, "in the city of Basel, [Switzerland,] where I live, there is a law that you may not use them in outdoor cafés. For the simple reason that they are an offense to the public."

The problem with the monobloc chair is not that it's ugly or anonymously made or extremely low cost, said MOMA curator Antonelli. "It's something ethical. It's made with lesser materials. It's not meant to last. It's a wasteful object." Earlier in her career, she worked with Giulio Castelli, the founder of Kartell. No one was a bigger booster of plastic furniture than he was. Over the years, she recalled with a laugh, "He assembled a collection of really, really ugly plastic chairs. To him they were so interesting because they showed how a material and its possibilities can bring out the best and the worst in people."

Despite all the bad furniture made from plastic, many designers still share their mid-twentieth-century predecessors' conviction that there are boundless possibilities to make something good. In recent years, the design world has been buzzing about the MYTO, a plastic chair designed by Konstantin Grcic and unveiled in 2007, not at a forum for design or furniture but at the plastic industry's biggest trade show, the triennial K Show in Düsseldorf. The choice was a nod to the chair's roots: the chemical manufacturer BASF had asked Grcic to come up with a design to promote its new ultrastrong polymer Ultradur. Grcic reached back to the Panton for inspiration and, working with BASF engineers, created the first plastic cantilever chair since the debut of that design icon. Grcic's take on that form is a hip, supple, springy zigzag of plastic with a perforated seat and back that Grcic hoped would evoke an animal's skin. It's so lithe it makes the Panton look stodgy. Thanks to BASF's new polymer and advances in processing technology, the chair has "an elegance that wasn't possible before," the MOMA's Antonelli said.

Photos of the MYTO have been posted on design blogs worldwide. The *New York Times* hailed it as one of the best ideas of 2007, the Museum of Modern Art added it to its permanent collection, and it was prominently displayed in a show about Grcic put on by the Art Institute of Chicago. *Times* design critic Alice Rawsthorn praised the MYTO for its "coolly angular shape" and for using "the minimum material possible." In this instance, that single shot of plastic is now a sign of eco-responsibility. In other words, the MYTO may be a monobloc, but it's one imbued with an ethic and intentionality that elevate it far above the $6.49 plastic patio chair. Grcic is just one of many contemporary designers enraptured by the possibilities of plastic. Kartell continues to be a major outlet for those designers' work. On a sunny spring afternoon, I went to visit the Kartell store in San Francisco, one of a hundred retail outlets the company has established in cities around the world.

The showroom felt like a cross between an art gallery and Ikea: plain white walls, recessed lighting, and every piece in the catalog out for display. The wares were grouped by color. There was a cluster of shiny reds: a sleek chair, a stool, a table with a lacy perforated top. Next to it was an orange array, followed by a gaggle of yellows. Across the room, a grouping of green chairs, tables, lamps, and vase-shaped pedestals shared a lighted platform. On the adjacent platform, much the same in cool shades of blue. Sunlight flooded through the floor-to-ceiling windows, making all those plastic surfaces extra glittery. It was like being inside a diamond — or, rather, a cubic zirconia. As someone accustomed to an earth-toned domestic world of pillowy upholstery and wood, I found the gleam and frank primary colors a little unsettling. Yet, I reminded myself, it wasn't as if my softer, more cushiony world was any less full of plastic. Like most modern furniture, my upholstered couches and chairs have polyurethane cushions, the covers are part polyester and sprayed with Teflon-like stain protectors, and many of my "wood" tables and bookshelves actually consist of fake wood veneers and epoxies over a partially plastic pressed-wood core.

Many of the pieces in the showroom were created by the legendary Philippe Starck, one of several prominent designers that the company began recruiting in the 1980s in an effort to upgrade its image. Starck's feelings about plastic echoed that earlier generation of designers: he loved the material for its democratic possibilities and because, unlike natural materials, it was the product of "human intelligence, so it fits with our human civilization." He also considered plastic environmentally preferable to using wood resources.

One of Starck's best-known designs is a beautiful chair called the Louis Ghost. Made of clear, hard polycarbonate plastic, the chair has an oval back, gracefully downturned arms, and curving legs — all taken from some classic and yet unspecified period in French history. Starck explained that he deliberately muddied the heritage: "I chose this icon to be the ghost of Louis 'I don't know what.'" Playful yet elegant, solid yet ethereal, the Louis Ghost has appeared in ads and in fashion magazines all over the world. Stylists have set it in starkly modernist rooms as well as in rooms filled with antiques. In either setting, it works.

Since its introduction in 2002, the chair has been one of Kartell's most popular pieces, selling many hundreds of thousands. This despite the fact that it costs four hundred dollars — which is not much for a traditional armchair but still quite a ways up the food chain from the unpedigreed monobloc. Somehow, the Louis Ghost has avoided both the pitfall of the avant-garde that kept the Panton chair from succeeding in the commercial marketplace and the stigma of cheapness that still bedevils the monobloc. I suspect the Louis Ghost has been so successful because it hits that sweet spot between cool and comfortable. Raymond Loewy, the grandee of twentieth-century industrial design, called it the MAYA principle — the most advanced yet most acceptable. The Louis Ghost takes full advantage of what plastic has to offer artistically without radically revising what we expect in a chair. The chair works because Starck accepted plastic on its own terms and plumbed its shiny, shallow waters for a genuine synthetic aesthetic.

I was curious to see how a monobloc chair would stack up against the Louis Ghost. So that afternoon I brought along one I had purchased at Home Depot, a model dubbed the Backgammon for no apparent reason. To my relief, the store manager was unfazed when I walked in with my Backgammon in tow. "Of course," he murmured smoothly when I explained I wanted to compare the two, as if it were an everyday request.

I took a couple of turns sitting in one and then the other. I can't say the Louis Ghost was much more comfortable than my chair from Home Depot. It was roomier than the Backgammon and provided more back support. But it was also so slippery that it was hard to comfortably settle in. The Backgammon dipped slightly when I plunked down in it. In truth, neither was a seat I'd want to spend a whole lot of time in. (Though I am sure the Louis Ghost would hold up better over time than my Backgammon. Not long after my visit to Kartell, my son leaned back in it too hard, and the spokes cracked.)

"It could be said, that when we design a chair, we make a society and city in miniature," the British architect Peter Smithson wrote. I look closely at the Louis Ghost and my Backgammon, trying to imagine the societies they evoke. One conjures a world of dazzling possibilities, the other a realm of cheap utility.

Looking at the two chairs together, I see a fair representation of the partner we've found in plastic: a Janus-faced companion who can rightly inspire both our deepest admiration and our strongest disgust.

I was outside the store on another day when a man and a woman came walking by arm in arm. They stopped for a moment to peer through the window.

"Look," the man said in a tone of utter incredulity, "it's *plastic* furniture."

"Yes," his companion answered, "but the designs are *gorgeous.*"

Flitting Through Plasticville

WHEN MY OLDER SON was born, a well-meaning friend—who had no children of her own—gave him a beautiful cherrywood rattle. It was smooth to the touch, safe to mouth, made a lovely plinking sound when shaken—and my son wanted nothing to do with it. He wanted the gaily colored set of plastic keys and, later, the squeaky vinyl bath book and, still later, the bright orange car with big blue wheels that made clicking sounds when it was pushed along the floor. Plastic is the medium of play today, so like most families with young children, we soon filled our house with enough junk to stock the midway at a state fair. We were forever tripping over remote-controlled cars, pulling plastic soldiers from between couch cushions, and cursing Lego when we stepped barefoot on the sharp-edged blocks in the middle of the night. My two sons accumulated an arsenal of plastic guns and Jedi swords. My daughter gathered a nursery of plastic baby dolls. (So much for our efforts to fight gender stereotyping.) For birthday parties, I stocked the goody bags with items from the catalog of the Oriental Trading Company, specialists in cheap plastic doodads: whistles, bouncy balls, squirt guns, glow sticks, all of which would invariably break or disappear minutes after the goody bags were distributed. It was only years later that I began to wonder: Where does this stuff *come* from?

My search for an answer to that question started one dreary winter day with a visit to the corporate headquarters of Wham-O, a company built on the wild, bouncy, springy, squishy, floaty possibilities presented by plastics. Wham-O introduced some of the most iconic toys of our age, from Hula-Hoops to Slip 'n Slides to its top-selling product, the Frisbee. Since the flying discs were introduced, in 1957, the company has sold more than a hundred million. Every American household surely has at least one; my family has somehow accumulated five, even though we almost never play with them.

This simple but ubiquitous toy offers an ideal window into the plastics industry, to the plants and processes that bond us ever closer with polymers by feeding our consumer desires. Plastics constitute the nation's third-largest manufacturing industry, behind only cars and steel. About one million Americans work directly in plastics. It's a sprawling industry that reaches into every sector of the economy, encompassing a few dozen petrochemical companies that create raw plastic polymers, thousands of equipment manufacturers and mold makers, and many thousands more processors that take raw plastics and fashion them into finished parts and products, such as toys.

Wham-O was started in Southern California, and its corporate headquarters are now in a modest one-story brick building in Emeryville, California, a sliver of a town wedged between Berkeley and Oakland. In the reception area, I was greeted by three big black-and-white photos of celebrities playing with Frisbees: a grinning Fred MacMurray (the classic TV dad from *My Three Sons*); the leads from *The Dukes of Hazzard;* and a distinctly pregubernatorial Arnold Schwarzenegger, in tight, skimpy shorts and a body-hugging T-shirt, spinning a disc on his finger. The prominence of the photos drives home how important the Frisbee remains to Wham-O even now, more than a half century after the toy's debut.

"It's really our bread and butter," explained David Waisblum, who at that point oversaw all aspects of the Frisbee brand, from manufacture to marketing. It was a dream job for Waisblum, a former stock-broker and self-confessed Frisbee freak who'd been an avid player of disc golf since he got out of high school. *Disc,* he explained, is the generic term for the toy. The name Frisbee is trademarked, so it can be used only for the flying discs that Wham-O makes. When I met him, Waisblum was in his early forties but looked much younger, partly because he was dressed in teen uniform: baggy jeans, sneakers, and a hoodie sweatshirt. Stocky, with shaggy brown hair, a goatee, and a mile-a-minute mouth, he reminded me of the actor Jack Black.

The company makes about thirty types of Frisbees and many were displayed on the wall in the conference room. It was a showcase of disc technology. Wham-O has found numerous ways to optimize discs: some glow in the dark; some have rims that make them easy for dogs to catch; some are heavy enough to slice through the blusters of a windy day. There are Frisbees specially engineered for the major disc sports: ultimate (a team game similar to football); disc golf (similar to regular golf except players aim for baskets, not holes); freestyle (spinning the discs and other discrobatics); and disc dog (just what it sounds like). Each demands a disc of a slightly different size, weight, and profile.

Then, of course, there are the basic recreational discs for your run-of-the-mill game of catch; they account for about half of all Frisbee sales. Waisblum wouldn't say how many Frisbees the company sold each year, but he claimed it was more than the annual sale of all base-balls, footballs, and soccer balls combined. I was surprised and skeptical, but to Waisblum it made perfect sense. "Balls are boring," he declared, then quoted another enthusiast who wrote that "when a ball dreams, it dreams it's a Frisbee."

In Frisbee genealogy, all descend from the original flying disc developed by the man Waisblum reverently referred to as "our inventor," Walter Frederick Morrison. In 1937, when he was a high-school student in Southern California, Morrison joined his girlfriend Lucille's family for Thanksgiving dinner, where he was introduced to the family game of "flipping" a big metal popcorn-pot lid. It was way more fun that just tossing around a ball, he decided. The next summer he and Lucille were flipping cake pans back and forth on the beach when a sunbather approached and asked if he could buy one. A business was born. The couple began peddling cake pans all along Southern California beaches, and Morrison started dreaming of ways to streamline and merchandise a better flying disc.

The business would have a long gestation. After serving as a fighter pilot in World War II, Morrison returned to Southern California, still enthralled by what he called "the Flittin' Disc idea." His stint in the U.S. Air Force had taught him something about what it takes to make things fly, and his cake-pan experience had convinced him he needed a material more pliable and less ding-prone than tin. Having seen how the new synthetic materials performed during the war, he thought to himself: *Plastic, that's just the ticket.* He spent several years trying out various designs and varieties of the recently introduced thermoplastics, hawking each new incarnation at county fairs. He and Lucille flitted the discs back and forth, mesmerizing onlookers with these new playthings that floated, dipped, skipped, and sailed in a repertoire of motion that balls rarely attained. The couple teased the crowds, claiming the discs were pulled along an invisible wire. The wire cost money, but anyone who bought one would get a disc for free!

In 1955, Morrison embarked on yet another redesign. This time he thickened and deepened the rim to increase its centrifugal force, and he added new details to give it more of a flying-saucer look, a nod to the public's growing fascination with UFOs. He added a small cupola on the top, where little green men might sit, along with the names of all the planets. He and Lucille, now married, dubbed it the Pluto Platter. It was their best flying disc yet. The discs were sold in plastic bags covered with references to the space theme, including the dubious instruction *Use bag for space helmet, if head fits.* One day, when Morrison was demonstrating Pluto Platters at a downtown Los Angeles parking lot, a man stepped out of the crowd and told him that the management at a local company had been thinking about marketing a flying disc. "It might be worthwhile to meet with the Boys at Wham-O," the man said.

The Boys were Rich Knerr and Arthur "Spud" Melin, high-school friends who had teamed up in 1948 to sell slingshots and sporting goods by mail order. Wham-O's early catalog was a modern parent's book of nightmares, filled with items guaranteed to put someone's eye out, if not remove a limb. There was the Malayan blowgun, with its "tempered steel hunting darts"; the throwing dagger, which was "balanced to stick"; and the cap pistol that "actually shoots peas, beans, tapioca, etc." As Knerr later recalled, "You couldn't buy those things just anywhere." Good as sales of such items were, by the 1950s, the pair could see there was an even brighter future in the business of toys.

The modern toy industry is in many ways the product of two major developments in the post–World War II era: the baby boom and the polymer boom. Though there had been plastic toys since the early days of celluloid—think of the kewpie doll—the convergence of those two broad trends sealed the marriage of plastic and play. After ramping up production for the war, the major manufacturers were swimming in supplies of the new thermoplastics, materials that could truly fulfill the British chemists' utopian dream of a world "where childish hands find nothing to break, no sharp edges or corners to cut or graze, no crevices to harbor dirt or germs." Thanks to the phenomenal postwar birthrate, there were millions of childish hands eager to play. During the peak years of the baby boom, annual toy sales leaped, from $84 million in 1940 to $1.25 billion in 1960. And an ever-increasing number of those toys were made of plastic: 40 percent by 1947. Today, plastics are a given in toy making; they're "like air," one manufacturer told me.

These cheap, lightweight, flexible materials vastly expanded play possibilities while raising profit margins. Fleshy vinyl permitted the manufacture of dolls that would "'feel' real as well as look real." Or not, as in the case of the impossibly curvaceous Barbie, who debuted in 1957. There were model cars, trains, and planes that had more detailing than wood or metal had ever allowed but that could still be sold for just a couple of bucks each. There were types of playthings never seen before, such as Silly Putty, developed by a scientist who was trying to create a synthetic rubber for the army in the early years of World War II. The military couldn't figure out what to do with it, but an entrepreneurial toy-store owner had an idea. And the SuperBall, which appeared in 1965 (and which Charles Eames considered one of the most elegant designs of the year). I remember how amazed my friends and I were by that little sphere of compressed black rubber (packed with so much energy that one early prototype tore apart the molding machine trying to get out). We'd spend recess bouncing the balls over one another, over the jungle gym, over the fences, over the roof until our exasperated teachers confiscated them.

The major plastics producers ran aggressive campaigns pushing plastics on Toyland. To promote its house brand of polystyrene, Styron, Dow Chemical invited manufacturers to submit toys made of the stuff for a corporate seal of approval. Those that passed muster were allowed to carry the Styron label, touting the material as "5 times tougher!" (Did anyone ask: "Than what?") Even after rejecting nearly half of the nineteen hundred submitted items, the company still issued well over ten million labels by the end of 1949. Companies also appealed directly to consumers: "Take it from the *Real* Santa Claus," a beaming Saint Nick declared in one 1948 ad in the *Saturday Evening Post*, "Toys of Monsanto Plastics bring Christmas Cheers."

But plastics' ascension was also the inevitable result of their sheer low-cost availability. In the early fifties, for instance, eight different chemical companies quickly built factories to start producing polyethylene, widely viewed as the most promising of the new plastics. Prices plunged to less than a dime a pound. The low cost stimulated scores of new applications, which absorbed supplies of the plastic, which in turn stimulated more production. All at once there was a host of new cheap toys, like dime-store cowboy-and-Indian sets and snap-together pop beads (which at one point were absorbing forty thousand pounds of polyethylene a month). Such boom-bust cycles have long driven the plastics industry, though throughout the wild ups and downs, for many decades it continued to grow, in some years and for some plastics at double-digit rates.

The most dramatic example of the ping-ponging relationship between supply and demand occurred when Phillips Petroleum tried to perfect production of a new semirigid variety of polyethylene. The manufacturing was tricky, and Phillips kept running into problems, turning out one unusable batch after another. Its warehouse filled with tons of off-spec, unsold plastic, a situation that threatened disaster until Wham-O came to the rescue in 1958. It started buying up the stockpiles to produce a new toy it had developed, the Hula-Hoop. After the singer Dinah Shore featured the spinning rings on her TV show, the hoops started flying off the shelves so fast that Wham-O couldn't keep up with orders. Tens of millions of hoops sold that first year, making short order of fifteen million pounds of material that until then Phillips hadn't been able to give away. Then, like so many fads, the craze for Hula-Hooping died as suddenly as it had taken off — and nearly took Wham-O down with it. Overnight, orders for Hula-Hoops dropped to zero. "We damn near went broke," Rich Knerr later recalled.

The Frisbee, however, has proved more enduring. And that's probably due to several things Wham-O did after securing the rights to Morrison's flying saucer. For one, Melin and Knerr rechristened Morrison's baby, picking a trademark that would distinguish their disc from the other Space Saucers, SkyPies, and Super Saucers then crowding the skies. *Frisbee* was a slight variation on the name used for a similar object in New England; since the 1930s, folks there had been tossing cake and pie tins from the Frisbie Pie Company and calling the sport Frisbieing.

Wham-O recognized that the Frisbee's longevity depended on its being seen as more than a novelty toy for playing catch. As Knerr and Melin had learned with the Hula-Hoop, even a best-selling toy can have a short shelf life. (Indeed, life in the toy market is so nasty and brutish that any toy that survives more than three seasons is considered a classic.) Sports, by contrast, have staying power and give rise to entire athletic ecosystems. Credit for nudging the Frisbee in that direction goes to a man known in disc circles as "Steady" Ed Headrick. After joining Wham-O in 1964, Headrick redesigned the Frisbee to make it more sport-worthy. He removed the goofy space references, broadened the saucer, and, to improve the aerodynamics, added concentric-circle ridges on the top, now known by discphiles as the "lines of Headrick." Such changes vastly improved the flight capabilities, making true disc sports possible for the first time.

Headrick himself invented disc golf, and he remained so passionate about the game and the disc that when he died, in 2002, he had his ashes molded into Frisbees. "He wanted all his friends to be able to throw him around," said Waisblum approvingly. "He wanted to come and rest on a roof somewhere, just out of reach, so he could bathe in the sun."

For all the advances in discs, the material used for the basic model has remained essentially unchanged since Morrison sold the company his Pluto Platter. It's the material that distinguishes a Wham-O Frisbee from a cheap knockoff (and cheap knockoffs are legion, since the disc-design patent has long since expired). Then, as now, it needed a material that was inexpensive, durable, and pliable, with the quality Waisblum called "givingness," which makes a disc pleasurable to catch and throw. Several plastics meet some of the specifications, but only one fulfills every item on Wham-O's wish list. That's polyethylene, the most commonly used polymer in the world and the one that, more than any other, molded the modern age of plastics.

Legend has it that one day John D. Rockefeller was looking out over one of his oil refineries and suddenly noticed flames flaring from some smokestacks. "What's that burning?" he asked, and someone explained that the company was burning off ethylene gas, a byproduct of the refining process. "I don't believe in wasting anything!" Rockefeller supposedly snapped. "Figure out something to do with it!" That something became polyethylene.

The story is almost certainly apocryphal. But I like it as myth, because it succinctly describes the origins of the modern petrochemical industry—a colossus grown from the principle that every hydrocarbon sucked from the ground can somehow help turn a profit. What is true is that Rockefeller's company Standard Oil was the first to figure out how to isolate the hydrocarbons in crude petroleum. That innovation helped give rise to the modern petrochemical companies that produce the raw, unprocessed polymers known as resins.

Most of today's major resin producers — Dow Chemical, DuPont, ExxonMobil, BASF, Total Petrochemical — have their roots in the early decades of the twentieth century, when petroleum and chemical industries began to develop alliances or form vertically integrated companies. Producers had begun to realize that there might be a use for the vast amounts of waste created in the processing of crude oil and natural gas and in the making of chemicals. Rather than being burned off as a worthless byproduct, ethylene could be retrieved and profitably deployed as a raw material for polymers. The growing reliance on fossil fuels helped drive the growth of the modern plastics industry, even though the production of plastics consumes a relatively modest amount of oil and natural gas. About 4 percent of global supplies of oil and gas is used as feedstock for plastics, and another 4 percent is used to actually produce them. Of course, an industry based on waste had one great advantage over rival industries: the low cost of its raw materials. The dregs of refining would always be cheaper than traditional materials such as wood or wool or iron.

With the rise of integrated petrochemical companies, the discovery and creation of new polymers became a more directed, rationalized process. Baekeland and Hyatt had gone hunting for synthetics that could replace natural materials in very specific applications, such as making billiard balls and electrical insulation. Starting in the 1920s and 1930s, industrial chemists became more interested in developing new polymers and only then finding ways to commercialize their discoveries. Plastics were moving into the economic driver's seat.

Polyethylene was discovered in 1933 by two chemists at Britain's Imperial Chemical Industries who were noodling around in the lab exploring how ethylene reacted under high pressure. In a series of experiments — including one that blew their reactor and much of their lab to smithereens — they found that with extreme pressure and the catalytic cajoling of benzaldehyde and a bit of oxygen, ethylene molecules hooked together into chains of stupendous length. The flakes of snow-white waxy stuff they found at the bottom of the reactor vessel were "so unlike the polymers known at the time . . . no one could envisage a use for it," one of the researchers recalled. Yet uses were soon found. Polyethylene turned out to be an able buffer of both high frequencies and high voltages. During World War II, the British took advantage of that dielectric property to develop the airborne radar systems that allowed them to detect and shoot down German fighter planes.

But polyethylene had other virtues. Lightweight, durable, "stiffer than steel, yet as soft as candle wax," chemically inert, endlessly re-moldable — this was a polymer that could and would be pressed into all kinds of service, from garbage bags to artificial hips, Tupperware to toys. In the 1950s, chemists developed improved ways of making polyethylene by using metal-containing compounds called metal-locenes rather than extreme pressure to catalyze the reactions that forged the polymer chain. The discovery allowed chemists to rear-range the daisy chains to create new variations on the polyethylene theme. High-density polyethylene, HDPE, a stiffer, semirigid mate-rial, became widely used for containers such as milk jugs and gro-cery bags. Low-density polyethylene, LDPE, and linear-low-density polyethylene, LLDPE, were flexible, stretchy materials, ideal for mak-ing filmy products such as plastic wraps and bags. Combined, they proved the perfect plastic for the basic recreational Frisbee.

Thanks to such versatility, polyethylene was the first plastic in the United States to sell more than one billion pounds a year, making it the first commodity plastic, meaning high volumes of it sold at low prices. Today, it's one of five families of commodity plastics that dom-inate the world's markets. Here are the other four: *Polyvinyl chloride,* also known as PVC and vinyl, has a remarkable shape-shifting ability: mixed with different chemicals, it can be made soft and flexible for things like shower curtains, hard and rigid for house siding and water pipes, or clear and filmy for packaging wrap. *Polypropylene,* a flex-ible, moisture-proof polymer, is used for monobloc chairs and food containers such as margarine and yogurt tubs. *Polystyrene,* which can be made as a hard, clear plastic, is often used in combs, hangers, and disposable cups, and it can be puffed up into Styrofoam. *Polyethylene terephthalate,* a type of polyester better known as PET, is a flexible, clear plastic used for soda and water bottles and as a spun fiber for making clothing and carpet.

Every day, of course, we encounter many other kinds of plastics. Altogether there are about twenty basic categories of polymers. They provide the foundation for tens of thousands of distinctive grades and varieties of plastics that are created by tweaking the essential characteristics of a given base polymer to make it more flexible, to increase its clarity, to improve its processing, or to confer any number of other desired properties. Still, the five basic families of commodity plastics make up the bulk of the market, accounting for about 75 percent of the roughly one hundred billion pounds of plastic produced and sold annually in the United States. Interestingly, all five date from the golden age of polymer innovation, the years bookending World War II. No significant new plastics have been introduced for decades. It's just too expensive and time-consuming to develop and bring an entirely new plastic to market. Polymer chemists today spend most of their time tinkering with and modifying existing base materials.

Of all the many plastics we rely on, polyethylene remains the favorite. For decades, it's constituted about a third of all plastics produced, largely because it is the polymer of choice in packaging. Indeed, according to calculations by Skidmore College chemist Raymond Giguere, the amount of polyethylene produced in America each year is nearly equal to the combined mass of every man, woman, and child living in the United States.

The company responsible for the biggest share of all that polyethylene is Dow Chemical. Which is why one spring day I found myself driving along Texas Highway 288 through a flat and treeless landscape toward the town of Freeport, where Dow has its biggest polyethylene plant. To see the origins of the plastic used in Frisbees, this was the place to go. Indeed, it would be the place to go to find the source of most of the plastics in my life: nearly all the raw plastic resins made in the United States come from petrochemical plants located along the fossil-fuel-rich Gulf Coast.

Dow arrived here in 1940, drawn not by petroleum but by the need for a new locale to sustain the historic heart of its business: extracting bromine and magnesium from brine to make chemicals. (The supply of brine in Midland, Michigan, where Herbert Dow founded the company in 1890, was almost tapped out. The Gulf of Mexico offered a near-limitless source.) But the region's deep stores of fossil fuels proved to be more important to the company as it expanded its production of polymers. When Dow announced its decision to buy eighty acres in what was then a tiny fishing village, the Freeport town fathers welcomed them with open arms. Ever since, the town and company have been wrapped in a tight embrace.

How tight became clear as I approached Freeport. For someone used to the strict zoning of San Francisco — where even a permit application to build a Burger King sets off a political brawl — I found the area that came into view a shock. One minute I was driving past low-rise apartment complexes, tree-shaded housing developments, restaurants, and shops, and then all of a sudden I was passing a vast industrial complex that stretched west as far as I could see, a dystopic skyline of otherworldly shapes, dun-colored towers, gigantic tanks, spires and silos and mazes of pipes. My hotel, advertised on the Internet as a local favorite spot for wedding parties, was right across the street.

Looking at a map later on, I saw that the town of 14,300 citizens was all but surrounded by petrochemistry. Dow operations cover five thousand acres in wide swaths to the northwest and east; a huge liquefied natural gas plant lies to the east also, bordering the beach; the salt-dome wells of the U.S. Strategic Petroleum Reserve sit on its southern edge. Scattered in between and around are the operations of other companies, including BASF, Conoco Phillips, and Rhodia. The only area without industry is to the west, where the municipal golf course is located. (It's also where, in 1994, chemicals leaching from a huge waste dump located on Dow property were found to be contaminating the ground water. A whole neighborhood had to be permanently evacuated.)

Dow's influence extends across the southern part of Brazoria County, named for the slow, muddy Brazos River that winds through it before emptying into the Gulf. Lake Jackson, the town north of Freeport, was literally built by Dow for the managers and engineers recruited to work at the new plant. Herbert Dow's grandson Alden, an architectural disciple of Frank Lloyd Wright, laid it out in the early 1940s. He designed much of the early housing and was responsible for the town's eccentric plan. Believing that it was nice not to know what lay ahead, he created an arrangement of insistently winding roads, all of them bearing names like Circle Way, This Way, That Way, and even Wrong Way.

Dow remains the biggest employer in the area and calculates that for every job the company directly provides, another seven are indirectly created. It pays more than $125 million in state and local taxes and donates more than $1.6 million each year to community projects large and small, from a maternity ward at the local hospital to new police radios. For more than fifty years, it's been the place where the county's blue-collar kids work when they graduate from high school, and where the white-collar sons and daughters hire on after getting their college degrees. Everyone knows someone who works at Dow. "If this place went away, the community would fold," said Tracie Copeland, the cheery public affairs representative who agreed to show me Plant B, one of the three production complexes that make up Dow's Freeport operations.

Plant B is a grid of fifty different plants, each one a mini-village devoted to the making of a particular plastic resin or chemical building block. We drove past facilities that make polypropylene, polystyrene, polycarbonate, and various epoxies, as well as the monomers — the starting chemicals — for polystyrene and polyurethane. Many encompass entire blocks. The streets have names out of the periodic table, such as Chlorine and Tin. Fat white pipes run along the ground or stretch overhead — the vital arteries of the complex, connecting it all. We passed a lone worker on an adult-sized trike, and I suddenly realized he was the only person I'd seen outdoors. More than five thousand people work at the plant, but as Copeland explained, this massive maze works is run from computerized control rooms; that's where all the people are.

The people may be scarce in this Twilight Zone–like ecosystem, but the wildlife isn't. Plant B is home to a large colony of migratory sea birds known as skimmers. A herd of longhorn cattle graze in one of its greenbelts, and big schools of tarpon and redfish live in the salt-water intake ponds, Copeland noted as she pulled over next to a long, slightly brackish rectangle of deep water. "You wouldn't think to look at it, but it's a great fishing hole," she said. Former Texas governor Ann Richards once stopped by here, she said, and "caught just a ton."

At the corner of Nickel and Glycol streets, we stopped to see where the production of polyethylene starts: the industrial glimmer-in-the-eye of what will eventually become a Frisbee. In front of us was one of Dow's two crackers, a block-long bank of gigantic furnaces that break down the hydrocarbon molecules in crude oil and natural gas. Either substance can serve as the base raw material for plastics. Dow uses natural gas, as have most American resin makers since the price of oil started rising in the 1970s. Today, about 70 percent of plastics made in America are derived from natural gas and 30 percent from oil; the reverse ratio holds true in Europe and Asia, where natural gas prices run higher than oil.

The cracking process uses a spectrum of temperatures and pressures to disassemble and reassemble those hydrocarbons into new arrangements of gases that will serve as the starting ingredients, the monomers, used in making plastics. When the carbon atoms form a ring of six, you get benzene, which is one of the bases for styrene, used to make polystyrene. A carbon quartet can become butadiene, a chemical used in making synthetic rubber and acrylonitrile butadiene styrene (ABS), the hard, shiny plastic used in Legos, cell phones, and other electronic devices. At another temperature, the carbons triple up, which can form propylene, the molecule used to make polypropylene. And in the highest hottest reaches of the cracker, where the temperature is cranked up past 750 degrees Celsius, two carbons can bond to form the gas ethylene, the starting molecule of polyethylene.

It was a quick drive from the cracker to a low, beige cinder-block building that serves as the nerve center for one of the operations dedicated to making the low-density polyethylene used in basic Frisbees. Our guide to the facility, John Johnson, was a fiftysomething barrel-chested mechanic in a blue jumpsuit who had worked at Dow since he got out of high school and who now oversaw maintenance of the plant. Production runs 24-7 and stops only for scheduled maintenance every eighteen months. "We run until something brings us down," he explained. In 2008, Hurricane Ike forced a closure, and it took two weeks to get the lines back up.

We followed him down a long hall, past a lab where polyethylene samples were tested for quality, to the control room, a space dominated by a long electronic board that looked like a digital subway map. But instead of tracking trains, this map tracked the flow of chemicals through the stations of transformation from various gases to liquid plastic. A man stood intently watching the board. "The board operator basically runs the plant," Johnson explained, then corrected himself. "The mod" — the computer system — "is running the plant, but he's the checks and balances for it. If something's out of line, he'll get the alarm and then come in and make adjustments." As if on cue, a bell sounded, and the man calmly punched a few buttons.

Before we could go outside to see the physical plant represented by the board, we were required to suit up. I pulled a blue mechanic's jumpsuit over my clothes, balanced an oversize hardhat on my head, slipped plastic safety glasses over my own glasses, stuffed earplugs in my ears, and pulled on thick leather gloves. All this to tour a facility in which, Johnson insisted, "you're safer than you are at home." Many of the chemicals used to make plastics, such as propylene, phenol, ethylene, chlorine, and benzene, are highly toxic. Decades ago, hazardous exposures were fairly common for plastics workers, but even critics agree the industry has improved its production processes, reducing the risks for its workers. "Dow has come a long way," Charles Singletary, the head of the local chapter of the operating engineers' union, told me. "We're not exposed like we used to be." Still, accidents happen. In 2006, a worker at the Freeport plant was exposed to chlorine gas during an accidental release. For some reason, his protective mask got pulled off, and the man inhaled some of the deadly gas.

According to Singletary, he called in the accident, finished his eleven-hour shift, went home, collapsed, and died.

Outside, Johnson showed us a bank of white pipes that carry the raw ingredients of polyethylene — ethylene, nitrogen, water, methane, and others — into and out of the plant. We followed the pipes, which arced overhead, into an enormous two-story shed filled with hissing and pumping machinery — the brute mechanics required to chemically crochet a new pattern of carbon and hydrogen atoms. Here, Johnson walked us past a series of tanks, compressors, and exchangers, explaining in great detail how the ethylene gas is repeatedly heated up and cooled down, squeezed under thousands of pounds per square inch of pressure, and then depressurized. After several cycles, more chemicals are piped into the mix: butane, isobutene, and propylene — the stuff that "makes the poly," Johnson shouted over the thundering sounds of production.

At the back of the second floor, he pulled open a door, then grabbed my arm as I instinctively started to walk through it. "You can't go in there. That's the reactor." This room represented the heart of the operation, the place where catalysts were added to the mix of chemicals to set off the big bang of the process: polymerization. Here was where the smaller individual molecules hooked themselves together into one magnificently giant molecule. I peeked through the door. I don't know what I was expecting to see — bubbling vats, steam-filled flasks. Instead, it was just a huge space filled with fat pipes looping up and down from floor to ceiling, like a gigantic intestine. I tried to imagine the molecules roller-coastering through the three-quarter-mile-long circuit of pipe, pulling closer and closer together, lining up, forming new bonds, gaining weight and mass until they dropped out of their airy gaseous state and pooled into a liquid resin.

I couldn't see any of that amazing transformation, of course. But as we returned to the ground floor and walked along the outer wall of the reactor chamber, I suddenly became aware that the atmosphere around us was subtly changed. The air had turned moist from all the hot water being fed into the reactor. The background noise shifted from a dull roar to a loud buzz, like a million lawn mowers. All at once I smelled plastic. My nostrils filled with that flat, featureless aroma you catch a whiff of when you chug the last drop from a plastic milk bottle or sniff a brand-new Frisbee.

The pipes around us were now flowing with liquid polyethylene. We followed them to another group of machines, where the liquid resin is cooled and molded into long spaghetti strands that are chopped into glossy rice-sized pellets, which are then spun dry. These pellets, also known as nurdles, are the coin of the realm in Plasticville, the form in which most plastics are traded and transported around the globe.

We watched a small hopper fill with fresh-baked white grains of polyethylene. I stuck my hand in; the pellets were still warm and so pleasing to the touch that I didn't want to pull my hand out. Johnson said the plant can make twenty-seven thousand to twenty-nine thousand pounds of pellets an hour, meaning that during the minute we stood watching, some four hundred to five hundred pounds of pellets tumbled by—roughly the combined weight of Johnson, Copeland, and me. It had taken scarcely sixty seconds to replicate our mass in plastic.

From here, the pellets are piped to nearby silos sitting alongside a pair of railroad lines. We climbed a flight of stairs into a shed that straddles the railroad tracks. From a catwalk we could look down. There were eight railcars lined up below us, each positioned precisely under a silo. Pellets poured like salt from a box of Morton's into a round opening at the top of the car. Each car holds 192,000 pounds of pellets, so the eight cars sitting below us would be carrying 1.5 million pounds of polyethylene. Some days just a single trainload is shipped out, and some days there are double shipments: sixteen cars — three million pounds of raw polyethylene — rolling out at five in the morning and five at night to factories around the United States and the world. Many will be loaded onto container vessels bound for China, where the pellets will be processed into products that we will then import. Dow, like other U.S. resin makers, has long supplied the plastics industries of the world.

That lopsided trade balance is changing, however. Historically, American and Western European companies have dominated the global industry, with the West supplying most of the nearly six hundred billion pounds of plastics now produced annually. But a seismic change is under way: the industry's center of gravity is shifting from the developed to the developing world, where production costs are lower, and demand and consumption are growing faster. China, India, Southeast Asia, and the Middle East have all been gearing up to produce their own raw plastic resins.

For oil-rich countries such as Saudi Arabia, Kuwait, and the United Arab Emirates, plastics are a natural next stop. Each has built new manufacturing complexes, and to jump-start those efforts, they've tried to ally with American petrochemical producers—always eager to get closer to their feedstock sources—to make various commodity plastics. The Saudi company SABIC, for example, purchased General Electric's storied plastics division in 2007. Thanks to such ventures, the Middle East's share of worldwide raw plastic production has increased fivefold since 1990, to 15 percent. Like the Chinese before them, the Saudis are trying to break into the value-added business of making finished plastic products. Those *Made in China* labels we're so used to seeing may be joined before long by products stamped *Made in Saudi Arabia.*

But those products may not necessarily be coming back to the United States or other developed economies. The United States, Europe, and Japan have long consumed the lion's share of the world's plastics, but as the romance with plastic goes global, experts believe the rest of the world is poised to quickly catch up. Per capita consumption of plastics in places like Africa, China, and India has shot up in recent years. A wide gap still exists—the world's average per capita consumption is still less than a third of the United States'. But that gap also suggests "a long trajectory of sustained growth of polymer production and demand in the developing world," as one recent forecast put it. Assuming the developing world falls for plastic as hard as Americans have, the rising demand, coupled with the growing population, will require plastics production to swell nearly fourfold by 2050, to almost two trillion pounds.

Who knows whether the pellets I saw streaming into the railcars were eventually made into Frisbees. There was something so abstract about the process that it was hard to connect it with any real-life plastic products. I wondered if Johnson felt a sense of ownership in things made of polyethylene, the way a mason might stop and admire a building where he had laid brick. "Absolutely," he said when I asked. "We sell a lot of stuff to S. C. Johnson to make Ziploc bags."

"So when you look at a Ziploc bag, do you feel proud?"

"Oh yeah, absolutely."

It's a long journey from a pellet to a Ziploc bag or a Frisbee. Along the way, the raw polyethylene passes through many different hands — compounders who mix in the needed additives; processors who make the finished product; brand owners who slap on a label; retailers who sell it. At every stop, the plastic gains in value. It costs Dow less than a penny to manufacture the 140 grams of polyethylene that go into a basic Frisbee. It costs the factory that makes Frisbees about twenty cents to buy that disc's worth of plastic, and it will spend another dollar or so on the costs of manufacturing and packaging the disc. Wham-O sells that disc to toy companies for three or four dollars. By the time that 140-gram Frisbee appears at my local toy store, it will sport a price tag of approximately eight dollars. The value of that hunk of polyethylene has risen by orders of magnitude. Still, as toys go, the Frisbee is a bargain.

Toy makers feel great pressure to keep their prices low, ideally under twenty dollars. Twenty bucks "is considered a magical price point because it is an 'ATM unit.' People think hard about breaking two of them," explained Danny Grossman, president of Wild Planet Toys and former president of the Toy Industry Association. Price points, of course, change with the times. With the 2008 recession, Grossman added, some stores began to look at fifteen dollars as the new twenty. Whatever the magic number is, the chief way the toy industry stays below it is by moving operations overseas. Welcome to China, where four out of five of the toys in the world are made.

Wham-O was late to join the procession of toy companies decamping from the United States. As long as Rich Knerr and Spud Melin owned the company, they kept it firmly planted in their Southern California home turf. The company had a factory in San Gabriel, and whatever toys weren't made there were farmed out to molders in and around Los Angeles. Indeed, until the 1970s, the whole area was full of plastics processors kept busy by the big toy companies. Every mold maker in Southern California "was doing Barbie legs and heads and parts," recalled one reporter who has long covered the industry. But then toy makers began moving production to Mexico, with Mattel and Kenner leading the way. (Toys were among the first of the major industries that use plastic to leave the United States. The continuing exodus of valuable end markets is a constant thorn in the side of the plastics industry, and one reason, along with the rising cost of natural gas, that it has been bleeding jobs for the past decade.)

Wham-O stayed put until Melin and Knerr sold the company in 1982. The new owners promptly moved production south of the border, and Frisbees were made by Mexican maquiladoras for the next two decades. In 2006, a Hong Kong–based toy company bought Wham-O — or what was left of it, for by then the brand was attached to only a handful of toys, including Frisbees, Hacky Sacks, and Hula-Hoops. To no one's surprise, the new Hong Kong owners moved Frisbee production to a vendor in China.

When I first asked if I could visit Wham-O's Chinese factory, the vice president for marketing and licensing turned me down, citing a need for secrecy that I normally associate with nuclear technology or Colonel Sanders's Original Recipe. Making a Frisbee "is not rocket science," he explained. "It's a very simple piece of injection-molded plastic. Any idiot can get a mold and make one. I don't want anyone in there unless he's from a government agency or Walmart or someone who absolutely needs to see it." Eventually, after much pleading on my part, he agreed to let me visit the factory, but with a proviso: I could not identify it or reveal where it was located on pain of a lawsuit. All I am allowed to divulge is that it is in Guangdong Province, in the Pearl River Delta, a place that's been described as the manufacturing center of the world.

For the past thirty years, this region just north of Hong Kong has been "the heart pumping China's emergence as a global economic power." As many as fifty thousand factories stud an area roughly the size of Missouri, turning out electronics, housewares, shoes, textiles, clocks, clothes, handbags, and countless other items, including 80 percent of the world's toys. To a large extent, what makes this beehive of productivity possible is plastic, the material used most often by all those industries. This is the most concentrated center for making plastic goods in China, if not the world, with some eighteen hundred factories and half a dozen huge wholesale resin markets where brokers peddle raw plastic pellets from around the globe. There are twice as many people working in plastics in that single province than in the entire U.S. plastics industry.

Before the economic crash of 2008, Guangdong's boomtowns drew tens of millions of migrant workers from the rural countryside and pulled in foreign investment at the incredible rate of nearly two billion dollars a month. Shipping containers left the region's busy harbors at the rate of one per second, around the clock, all year round, journalist James Fallows calculated. If the region were a country, at that time it would have boasted the world's eleventh-largest economy.

It's also one of the most densely populated places on earth, with an estimated population of forty-five to sixty million. (No one is quite sure, because of all the migrant workers.) It was hard for me to appreciate the implication of such numbers until I got on the train from Hong Kong and reached the district's first major city, Shenzhen. All I could see were complexes of skyscrapers stretching out in every direction. It looked as if multiple copies of midtown Manhattan had been cut and pasted under the dull gray skies. (The smog overlying the province is so thick and persistent that it killed off the region's centuries-old silk industry. By the 1990s, the silkworms just couldn't be kept alive.) The only breaks in the skyscraping came when I passed the big block-shaped factories, which for some reason are almost invariably five stories high.

Thirty years ago, Shenzhen was a sleepy fishing town of about seventy thousand people. Now it has a population of about eight million. "It changes every day," my translator Matthew Wang later told me. He spent a few years working in factories there. For him, it was a lonely time. "In this city, you need to keep moving, moving. Nothing is stable. Accommodations, jobs, friends, everything. That's why economically it's good, but it's not a good place to live. My wife says I would have gone crazy if I'd stayed here."

Matthew, nearing forty at that time, embodied this feverish pace of change. He was the son of peasants. His father was briefly jailed during the Cultural Revolution. Matthew was raised in a farming village, drank water hauled in a wooden bucket from a well, and did his schoolwork by the light of kerosene lamps. But he did well in school and mastered English, and now he was a player, albeit a small one, in the global economy. He followed international affairs on the Internet (to the extent Chinese censors allowed) and made his living as a translator and fixer for foreigners with business in Guangdong, like me. One day as we were driving to an interview, his ever-present cell phone rang with a call from an Australian client who wanted him to make arrangements for a shipment of shoes to Sydney.

This steroidal push into the twenty-first century was all the more surreal for the contrary images that kept popping up, the reminders that this sheen of development and prosperity reached only so deep. Bicyclists pedaled along the sides of traffic-choked six-lane highways. Peasants in straw hats hand-hoed little pockets of farmland on the outskirts of cities. Towers under construction were framed by bamboo scaffolding. Drying laundry hung from the balconies and windows of every high-rise building.

Though Guangdong has been a locus for international trade on and off since 200 B.C., this current gold rush of foreign investment began in 1979 when Premier Deng Zhou Peng announced his open-door policy. Under a series of economic reforms, the government established "special economic zones" in the cities of Dongguan, Shenzhen, Guangzhou, and Foshan. Each was granted special tax benefits that made it attractive to foreign investors, especially those based in nearby Hong Kong, which was still a British colony.

By then Hong Kong had a strong plastics-processing industry, geared heavily toward export. As in the United States, Hong Kong plastics manufacturers had started in the 1940s with simple objects like combs, then moved on to toys, and by the 1980s they were producing for the more lucrative end markets, such as computers, cars, and medical devices. But toys remained a mainstay export. Enticed through Deng's open door, plastics processors and toy makers began migrating to mainland China, where the rents were cheaper and the labor far more abundant. To this day, most of the toy factories in Guangdong Province have Hong Kong or Taiwanese owners.

The path that brought the owner of the Frisbee factory, Dennis Wong, to the region is typical. Born and raised in Hong Kong, he studied polymer engineering at Hong Kong Polytechnic and cut his teeth in the industry working for Union Carbide's Hong Kong outpost. Hong Kong's homegrown plastic industry was still young, he recalled. "All the information was from America. All the plastic molding technology, all the equipment and how to mold it, how to machine it, how to make good plastic products was introduced from America." When he and his wife started their company, in 1983, they began by making simple, practical items, such as flashlights and refrigerator magnets. The company developed a reputation for doing good work. One day a toy company asked Dennis if he could manufacture its novelty pens. Soon Dennis was in the toy-manufacturing business.

In 1987, he built a factory in Guangdong Province, in what was then a remote spot in the countryside surrounded by rice paddies. It took a taxi two hours to reach the factory from the local train station, and invariably the driver would get lost. Now a busy thoroughfare runs past the front gate, and it's surrounded by a bustling neighborhood of shops, hotels, apartment buildings, and other factories. Though Dennis comes to the factory nearly every day, he and his family continue to live in Hong Kong, a ninety-minute drive away. They all work in the business. The company employs about a thousand people, which is small by Guangdong standards. Still, it enjoys a strong reputation.

Most of the company's work is dedicated to manufacturing other companies' branded products, such as the Frisbee, as well as anonymous tchotchkes such as key chains, light-up pens, and pedometers. But like many Chinese processors today, Dennis's daughter Ada has higher ambitions. She is hoping the company can eventually start producing its own brand of toys; that's where the future is. To that end, her business card reads *Product Innovation Manager*. She's a friendly, slender woman in her early thirties with chin-length hair and delicate features. She speaks impeccable English. She drove up from Hong Kong to guide me through the factory on a broiling-hot March day.

Ada had promised to show me the production process from start to finish. Accordingly, our first stop was a small room off the main production area, where the raw resins are mixed into the custom blends that Wham-O requires for its Frisbees. Bags of clean white pellets were stacked against the wall, and I spotted the label of ExxonMobil among them. (The factory almost exclusively uses resins from overseas — the United States, Taiwan, Mexico, the Middle East — because, Dennis later explained, Chinese resins are not reliable; the quality can vary from batch to batch, which can affect processing. Despite China's powerhouse status as a producer of plastic products, it still imports most of the resins it uses, though construction of new resin plants will start changing that equation.) Those pellets — a mix of high-density and low-density polyethylene — are blended in a barrel with grains of pigment and softening agents in proportions that Wham-O has prescribed. The raw materials are then ready to be made into Frisbees.

Out in the clanging, whirring din of the main factory floor were six injection-molding machines — each about the length of a limousine — devoted to molding Frisbees. (More were going full-bore in another building.) I stopped by one and watched the process. It reminded me of a giant Play-Doh Fun Factory. A funnel-shaped hopper sitting on top of the machine was filled with a blend of pellets and white pigment. Every so often, the hopper released a batch into a long horizontal barrel, where they were immediately heated to 400 degrees Fahrenheit. As the plastic melted, a long screw pushed it through the barrel into a Frisbee-shaped cavity formed by molds that clamped together with a pressure of more than two hundred tons per square inch. The mold was chilled so the plastic started hardening as soon as it reached the cavity. All this took fifty-five seconds. At that point, the front of the mold pulled away from the back, and a woman sitting next to the machine opened a small glass door and plucked out a shiny white 140-gram Frisbee. She carefully inspected it for flaws and snipped off any trailing filaments of plastic as well as the sprue, the little tab indicating where the liquid plastic ran into the mold. By then, another fresh Frisbee was ready to be pulled from the mold. Once the Frisbee cooled down, it was placed onto a rack with hundreds of other blank discs awaiting decoration. One disc she pulled out had a little smudge of red on top, residue from a previous production run. She razored out the offending spot to avoid further contamination and threw the disc onto a pile of rejects that would be remelted and remolded into new Frisbees.

This may not be rocket science, but it's more complicated than it looks. It's taken the company much trial and error to ensure the discs contain just the right blend of materials, that they come out at the proper weight, and that they don't deform while cooling, said Ada. Indeed, the company spent a considerable amount of money upgrading its machines, buying new equipment, and machining new molds in order to make Frisbees. What has made it worthwhile? I asked. "Quantity," Ada answered, without hesitation. The company was producing over one million discs a year.

Actually, it was making one million discs in the space of about four months. Because the peak season for Frisbees is the summer, the factory hums with disc-making from January to April only. After that, the machinery is refitted with different molds to produce other toys for the U.S. Christmas rush. The seasonality of toy making means many Guangdong toy factories fall idle and lay off workers for several months of the year. Dennis has been both savvy enough and lucky enough to keep his company going full-speed year-round.

Ada took me upstairs to the area where the discs are decorated. A pair of women were sitting in front of hot-stamping machines, bought specifically for Frisbee production. One fit a faceless disc into the machine and — *whoosh* — the top was stamped in black with the image of what looked like an octopus surrounded by a ring with the words *All Sport* and *140 gram*. She passed the Frisbee to her partner, who fit it precisely into place on her machine, which then embossed in silver a swooshing pattern of circles and the logo FRISBEE DISC. Nearby were dozens of racks filled with bright blue, yellow, orange, and white hot-off-the-presses Frisbees.

Ada had mentioned that she had about a hundred employees working on Frisbees. So far, I had counted a dozen at best. It turned out that the job requiring the most manpower — or, more accurately, womanpower, since almost every worker I saw was young and female — was the packaging of the discs. We walked up a flight of stairs and entered a big open room, where two long production lines were devoted to making the discs retail ready. Young women sat alongside conveyor belts, each bent to the single task that she would repeat hundreds of times a day for the duration of the Frisbee production run, whether it was arranging six discs in display boxes, fitting labels onto the discs' undersides, adding production codes to the labels, sealing clamshell casings, or packing discs in big cardboard boxes stamped *Made in China*. The only automation was the moving conveyor belts. The space was spacious and well ventilated, but even with all the windows open, it was still brutally hot, and it wasn't even summer yet. There was no air-conditioning.

The leader of one of the production lines was Huang Min Long, a solidly built woman in a T-shirt and jeans, her hair pulled back under a blue cap. Like most of the workers in the factory — and other factories throughout Guangdong — she was a migrant. She *went out*, the term commonly used by migrant workers, three years before from Guangxi, a region hundreds of miles to the west, leaving behind two children. She only got to see them once a year, when she went home for the spring-festival holidays. The rest of the year, she lived in the company dormitory, sharing a room with as many as nine other women. The room I saw was a cramped space filled with bunk beds. A single fan hung from the ceiling next to a fluorescent light. One bed was stacked with small lockers where the women could stow their private belongings, another with their suitcases. Each bunk had a sheet hanging across it — the only measure of privacy. Plastic tubs for doing washing were piled out front, and a communal bathroom was down the breezeway, near the canteen where they took all their meals.

The migrant's life is a difficult one. Ada was not willing to provide details or let Huang talk about how much she and other employees were paid or the hours they worked. But toy-factory workers put in notoriously long hours for desperately low wages. At that time, the average worker at Mattel's plant in Guanyao (also in Guangdong Province) made $175 a month for a sixty-hour workweek. She had to pay for her dormitory housing and meals out of those wages. Recent labor-law reforms may improve migrant workers' lots somewhat. But they will have a long way to go. The watchdog group China Labor Watch reported in 2007 that conditions in many toy factories are "devastatingly brutal," marked by "long hours, unsafe workplaces and restricted freedom of association." In some factories during peak season, workers are forced to put in ten- to fourteen-hour days for weeks, without a single day off. According to the report, factories impose illegal fines and penalties that cut further into the employees' meager pay. The group lays a good share of the blame not on the factory owners but on the multinational toy companies and big-box retailers that insist on being able to sell toys for under twenty dollars each. Cheap toys have their price: "In order to maintain even modest profits, many of these factories have no choice but to accept toy companies' low prices," the group noted. "Sadly, workers' salaries and general treatment are the only flexible factor of production . . ."

This may not be the case with Wham-O or the Frisbee factory. My brief tour wasn't enough to fairly assess conditions there. It looked clean and safe, and though the dormitory and open-air canteen seemed depressing by American standards, my translator assured me that he had seen worse. While migrant workers typically change jobs frequently, Ada told me the workers at her company tended to stay put. "I don't know why," she said, "but our factory people stay for a longer period of time. Some of our workers [have been] working here for twenty years."

Virtually all of the factory's products are destined to go overseas. That export orientation built China's franchise in Plasticville, but in the absence of strong domestic markets, it leaves China's toy makers vulnerable to world events, such as the epidemic of toy recalls that took place in 2007. The discovery of lead paint and other safety hazards forced American toy companies to recall more than twenty-five million Chinese-made toys that year. Those recalls, coupled with the global recession starting in 2008, knocked the Chinese industry to its knees. By some estimates more than five thousand toy companies — not only in Guangdong — closed between mid-2007 and early 2009. Meanwhile, an unknown number of others have been relocating operations to less expensive parts of China or to cheaper countries such as Vietnam, following the same well-traveled path that initially brought the toy industry to Guangdong. The provincial authorities are apparently happy to see them go, eager to replace the light manufacturing that ignited China's roaring economic engine with higher-value, higher-tech industries. These will also rely heavily on plastics.

For all his success, Dennis proved to be vulnerable. Six months after I visited his factory, new owners bought Wham-O and decided to cancel his contract, with little regard for the hefty capital investment his company had made in order to produce Frisbees. The new owner, Marvel Manufacturing, has its own manufacturing facilities in China, as well as in Mexico and the United States. It announced it was returning Frisbee production to the United States, though as of mid-2010, the vast majority of discs were still being made at its factory in China.

Loss of the Frisbee contract was a disappointment, but that's the way business goes, Ada told me when I contacted her after the sale. She said the company had replaced the work for Wham-O by moving into a new market niche: musical greeting cards. It gives the company an entrée into electronics, which is a step up from toys. "Toys are not that stable," she said. The tune-playing cards are "more mass market."

As it turns out, greeting cards that chirp a toneless "Happy Birthday" are more comprehensible to Ada and her employees than a high-flying platter of polyethylene.

"Is Frisbee very famous in the U.S.?" Ada had shyly asked me at one point as we walked through the factory.

"Sure," I told her, "it's very famous. Everyone has had a Frisbee at one time or another."

To Ada and others in the factory, that popularity was a mystery. "In Hong Kong it is not popular. So we are just thinking, why are so many people ordering Frisbees?"

"So people don't play it here?" I asked.

"Oh, no, only once in a while we play it on beaches."

Huang, the worker I briefly spoke with, was equally baffled by the toy that she spent months packing into boxes bound for overseas destinations. I asked my translator Matthew Wang to ask her what she thought people did with Frisbees.

"She knows it's used at the beach."

"Has she ever played with a Frisbee?" I asked.

"No," Matthew translated. "She's never been to a beach."

"Humans Are Just a Little Plastic Now"

BABY GIRL AMY* was born in April of 2010, four months early and weighing not much more than two Big Macs. She was whisked straight from the delivery room to the neonatal intensive care unit (NICU) at Children's National Medical Center in Washington, D.C.

When I saw her two days later in the NICU, I couldn't help but gasp. She was perfectly formed but still seemed so unfinished, with fingers like tiny spring twigs, and skin as translucent as a new leaf. She was in an enclosed clear-plastic incubator, connected to a jumble of tubing. Foam pads covered her delicate eyes to protect them from the special UV lights used to prevent jaundice. Aside from the nest of soft blankets on which she lay, she was entirely surrounded by plastic.

Neglect and negligence had hurried her into the world. Her mother had had no prenatal care. She had a drug problem. At the time she went into early labor, she was high on angel dust. She'd been carrying two babies, but Amy's twin was stillborn, and Amy's chances were precarious. "We weren't even expecting her to make it this long,"

---

* This is not her real name.

said the nurse caring for her. Billie Short, the doctor in charge of the NICU, gave her a 40 percent chance of survival. The fact that Amy had survived her first few days and could even go on to make it was in a sense a victory of polymer technology. Neonatology, like much of modern medicine, owes a huge debt to the advent of plastics, in ways both spectacular and mundane.

Polymers made possible most of today's medical marvels. Dutch physician Willem Kolff, driven by a conviction that "what God can grow, Man can make," scrounged sheets of cellophane and other materials in Nazi-occupied Holland to perfect his kidney-dialysis machine. Today, plastic pacemakers keep faulty hearts pumping, and synthetic veins and arteries keep blood flowing. We replace our worn-out hips and knees with plastic ones. Plastic scaffolding is used to grow new skin and tissues; plastic implants change our shapes; and plastic surgery is no longer just a metaphor.

Plastics are in the housing and components of sophisticated imaging devices. They also supply the essential everyday equipment of medicine, from bedpans and bandages to the single-use gloves and syringes that first appeared in the 1950s but became utterly indispensable in the wake of AIDS. With plastics, hospitals could shift from equipment that had to be laboriously sterilized to blister-packed disposables, which improved in-house safety, significantly lowered costs, and made it possible for more patients to be cared for at home.

Sizewise, medicine is a small end market, consuming less than 10 percent of all polymers produced in the United States — peanuts compared to packaging (33 percent), consumer products (20 percent), and building and construction (17 percent). But it's a strong, recession-proof market and one that has provided the industry enormous PR value. Medicine has long been plastic's indisputable good-news story, the showcase of polymers' benefits. For one recent public relations campaign, the American Chemistry Council featured a photo of a newborn in a plastic incubator.

Plastics are indispensable in neonatology, agreed Dr. Billie Short as we toured the fifty-four-bed NICU at Children's National, where preemies like Amy may spend the first several weeks or even months of their lives. Short is chief of neonatology at George Washington University, and as we stood by Amy's bedside, she described how plastics enable the care of infants this fragile. Reaching her hands through the pair of portholes in the sides of Amy's incubator, she pointed out the quartet of clear, exquisitely thin tubes that were delivering nourishment and medicines to Amy from the several plastic bags hanging on a nearby rig. One was inserted into a vein in her head to provide fluids; another, delivering antibiotics, tapped into a vein in her arm, which itself was scarcely wider than the pen I was using to take notes. Two catheters ran into the stump of her umbilical cord, one to feed nutrients into a vein and another connected to an artery so that the nurses could monitor Amy's fluctuating blood pressure and the levels of oxygen in her blood. A respiratory tube threaded down her throat was attached to a plastic-encased machine that helped her breathe. All the tubing was soft and supple enough to slide through her delicate body without tearing anything. Meanwhile, that enclosed plastic incubator maintained a carefully calibrated humidity and warmth (preemies like Amy don't have the layers of skin and fat needed to sustain body temperature). This kind of equipment is one of several factors that have helped raise the survival rates of premature babies over the past forty years.

I watched her chest rise and fall as rapidly as a sparrow's. Every so often, an involuntary tremor would ripple across her tiny body, as if she were shuddering over whatever rude force in the universe had pulled her from the dark coziness of her mother's womb into this synthetic approximation.

How long will she be hooked up to all this? I asked Short, indicating the intravenous tubing.

"Oh, for weeks," said Short. After that, once she became stable enough, Amy would receive her nourishment through feeding tubes.

Neonatology is a relatively new medical specialty. The first NICU was set up in 1965. That the field has blossomed in the age of polymers is probably not a coincidence, given the challenges of treating babies with hair-thin veins and tissue-paper skin. Still, until the 1980s, most of the intravenous fluids used in NICUs came in glass bottles. Short remembers the worry and inconvenience of those bottles falling and breaking. At first, said Short, the move to plastic seemed a tremendous advance. "We all thought plastics were inert, safe. We didn't have to worry about it. Then as the research came out, it became more and more evident we needed to pay attention."

And here, Short hit on the central paradox of plastic in medicine: in the act of healing, it may also do harm. Research now suggests that the same bags and tubes that deliver medicines and nourishment to these most vulnerable children also deliver chemicals that could damage their health years from now. The vinyl plastic typically used in IV bags and tubing contains a softening chemical that can block production of testosterone and other hormones. This chemical, called a phthalate (pronounced *tha-late*), doesn't act the way familiar environmental villains such as mercury and asbestos do; for those substances, there's a direct connection between exposure and easily recognizable subsequent harm, such as cancer or a birth defect or death. Phthalates leave tracks along more complex, convoluted routes. That's because they play havoc with the body's endocrine system — the intricate, self-regulating choreography of hormones that dictate how an individual develops, reproduces, ages, fights disease, and even behaves. Phthalates are not the only chemicals used in common plastics that have disruptive effects. By mimicking or blocking or suppressing production of hormones such as testosterone and estrogen, these various chemicals may produce subtle, long-term effects that don't show up for years or appear only in our offspring. They may make us more vulnerable to asthma, diabetes, obesity, heart disease, infertility, and attention deficit disorder, to name just a few of the health problems that have been linked to various of these chemicals in animal studies and epidemiological surveys. And some of these substances may do their damage even at minute concentrations we never considered worrisome.

Just as plastics changed the essential texture of modern life, so they are altering the basic chemistry of our bodies, betraying the trust we placed in them. All of us, even newborns, now carry traces in our systems of phthalates and other synthetic substances, such as fire retardants, stain repellants, solvents, metals, waterproofing agents, and bactericides. Though the chemicals surely don't belong there, the actual threats to human health are still uncertain. As different as my life is from baby Amy's, I can't help but see similarities. In the age of plastics, we are all incubator babies, inescapably tied to polymers, facing a world of new risks.

Few objects speak to the medical impact of plastics — the benefits and the risks, and the complexities of balancing the two — as well as the plastic IV bag with its snakelike tubing.

This staple of health care was introduced in the years following World War II by Carl Walter, a surgeon and professor at Harvard Medical School. Like many surgeons, Walter had a mechanical mind and a gift for invention. In the late 1940s, he turned that talent to the problems plaguing blood collection and storage. The whole idea of blood banking was still new. Walter himself had established one of the first blood banks just a decade before, locating it in an obscure basement room at Harvard to avoid arousing the ire of university trustees who considered it "unethical and immoral" to collect and use human blood. Blood banks of that era faced significant problems. Donors' blood was drawn through rubber tubes into rubber-stoppered glass bottles, a process that often damaged the red blood cells and allowed bacteria and air bubbles to get in. Searching for a better system, Walter hit upon the idea of using one of the most exciting of the new thermoplastics, polyvinyl chloride, better known as PVC, or vinyl.

PVC is a unique polymer. Unlike other plastics, PVC has chlorine as one of its chief ingredients, a greenish gas that is derived from a salt (sodium chloride). To make PVC, the chlorine is mixed with hydrocarbons to form the monomer vinyl chloride, which is then polymerized, resulting in a fine-grained white powder.

This unusual chemistry is PVC's greatest strength, but also its greatest problem — the reason that industry sings its praises and that environmentalists call it Satan's resin. The chlorine base makes PVC chemically stable, fire resistant, waterproof, and cheap (since less oil

or gas is needed to produce the molecule). It also makes PVC hazardous to manufacture and a nightmare to dispose of, because when in-
cinerated it releases dioxins and furans, two of the most carcinogenic
compounds in existence.

PVC is also an unusually polyamorous molecule; it's amenable to
hooking up with a host of other chemicals that can lend the resin an
extraordinary array of properties. Indeed, without additives, PVC is
so brittle that it is virtually useless. But combined with other chemicals, it can be "converted into an almost limitless range of applications," as its pitchman the Vinyl Institute boasted. It can be made
into the stiff planks used to side houses, the strong pipes that carry
water, the insulating coating of electrical wires, the squeezable arms
of a doll, the soft drapes of a shower curtain, the fleshlike texture of
a dildo. Such versatility has made PVC one of the top-selling plastics
in the world and a frequent choice for makers of medical devices. But
owing to the resin's dependence on additives, it has come under fire.

The material that caught Carl Walter's eye was a particular variety
of vinyl, known as plasticized PVC, in which the plastic is made soft
and pliable through the addition of a clear, oily liquid called di(2-
ethylhexyl) phthalate, or DEHP, a member of the phthalate family.
Phthalates have become so ubiquitous in consumer and industrial
products that manufacturers make nearly half a billion pounds of
them each year. They're used as plasticizers, lubricants, and solvents.
You'll find phthalates in anything made of soft vinyl. But you'll find
phthalates in other types of plastic and other materials too, in food
packaging and food-processing equipment, in construction materials, clothing, household furnishings, wallpaper, toys, personal-care
products such as cosmetics, shampoos, and perfumes; adhesives,
insecticides, waxes, inks, varnishes, lacquers, coatings, and paints.
They're even used in the time-release coating for medications and
nutritional supplements. There are about twenty-five different types
of phthalates, but only about half a dozen are widely used. Of those,
DEHP is one of the most popular, especially for medical uses. And
for that, we can thank Carl Walter.

For Walter's purposes, plasticized PVC seemed ideal: it was durable, flexible, and, unlike glass, didn't seem to damage red blood cells. It permitted CO_2 in the blood to dissipate and oxygen to disperse, which was also good for the blood cells. As far as he could tell, the material was completely inert. To persuade colleagues of its virtues, he brought a blood-filled bag to a meeting, dropped it on the floor, and then stepped on it. The bag held, which in itself was a huge advantage over breakable glass. But the real advance was that the bag could be connected to other bags to form a secure, sterile system through which blood could be separated into its various component parts — red blood cells, plasma, and platelets.

The new technology revolutionized the way blood was collected and used. For the first time it was possible to safely separate and store blood components. Instead of giving a patient whole blood, a physician could administer only the parts a person needed. A single unit of blood could now be stretched to help three different people.

The U.S. Army employed the new bags during the Korean War, and they proved a vastly safer and more reliable way to treat injured soldiers in the field. Medics could squeeze the bags to get the contents to flow faster. Glass bottles, on the other hand, worked by gravity alone; they had to be hoisted high above the patient, creating a target for enemy fire. By the mid-1960s, PVC blood bags were standard in civilian blood banks and hospitals. Walter's innovation also began rippling through the field of intravenous therapy. Over the decades, medical suppliers gradually swapped out glass for PVC to hold a wide range of fluids that were delivered intravenously, from saline to drugs to nutritional supplements.

One of PVC's big selling points was its presumed chemical stability. As *Modern Plastics* pointed out in a 1951 article entitled "Why Doctors Are Using More Plastics": "any substance that comes in contact with human tissue . . . must be chemically inert and non-toxic," as well as compatible with human tissue and not absorbable. PVC was one of the plastics that seemed to fit that bill. But in the late 1960s and early 1970s, a series of revelations began chipping away at that presumption of innocence.

First, there was the discovery that vinyl chloride gas—the key chemical used to make PVC—was far more dangerous than had been supposed. Doctors at B. F. Goodrich's PVC plant in Louisville, Kentucky, discovered in 1964 that workers there were developing acroosteolysis, a systemic condition that caused skin lesions, circulatory problems, and deformation of finger bones. Then, in the early '70s, European researchers found evidence that vinyl chloride was carcinogenic. As David Rosner and Gerald Markowitz detailed in their exposé *Deceit and Denial: The Deadly Politics of Industrial Pollution,* the vinyl industry initially suppressed results of studies that showed that low levels of vinyl chloride caused liver cancer in rats. But the cover-up was revealed in 1974 when four workers at the Goodrich plant died from that same rare liver cancer, angiosarcoma. A writer for *Rolling Stone* likened the Louisville factory to "a plastic coffin."

Frightening as the Louisville revelations were, they described a familiar environmental hazard, one stemming from dangerous workplace conditions and largely confined to the factory floor. If vinyl chloride caused cancer in PVC workers, then plant conditions would have to be changed so workers were no longer in danger. And indeed, after a contentious regulatory and legal battle, the newly formed Occupational Safety and Health Administration set a strict threshold that dramatically limited workers' exposure to the chemical. (The ruling dropped acceptable levels from five hundred parts per million to one part per million.) Industry howled, claiming that the cost of complying with the new standard would run as high as $90 billion, but in the end, making plants safer cost just a fraction of that, only $278 million. Since then, no cases of angiosarcoma have been reported among vinyl workers.

While the vinyl chloride scandal was unfolding, another line of research was pointing to a more insidious—and also more uncertain—risk: that the chemicals added to PVC were leaching out of widely used products.

Johns Hopkins University toxicologists Robert Rubin and Rudolph Jaeger stumbled onto that discovery during a 1969 experiment involving rat livers. The livers were being perfused with blood from PVC blood bags and tubes when it became apparent that some unknown compound was confounding the experiment. Rubin asked Jaeger, then his graduate student, to figure out what the mystery compound was. Jaeger discovered it was DEHP, the plasticizing chemical added to the vinyl that was used to make blood bags and tubing. As he would soon find, those bags could be as much as 40 percent DEHP by weight; the tubing could be as much as 80 percent. The additive is not atomically bonded to the molecular daisy chain that makes up PVC, which means it can migrate out, especially in the presence of blood or fatty substances.

Neither Rubin nor Jaeger knew if DEHP was toxic, but they were surprised and a little alarmed by follow-up studies in which they found traces of the chemical in stored blood as well as in the tissues of people who had undergone blood transfusions. When Jaeger submitted their paper documenting those findings to the prestigious journal *Science,* the editor rejected it, telling him that unless the chemical was demonstrably toxic, it didn't really matter if it was getting into people. Jaeger boldly called the editor back and persuaded him to change his mind. As he recalled, "I said, 'Look at the extent to which phthalates and PVC plastic are used in society. It's worth publishing because scientists need to be informed about the ubiquitous nature of extracts from plastics.'"

Not long after, a chemist at the National Heart and Lung Institute reported that he too had found residues of DEHP and other phthalates in blood samples taken from a sample population of one hundred people. But these were not people who had been exposed through work or who had undergone blood transfusions; they were simply consumers of plastic goods, folks who might have been exposed to phthalates from any of thousands of everyday products, from cars to toys, wallpaper to wiring. Reporting the findings in 1972, the *Washington Post* declared, "Humans are just a little plastic now."

What, exactly, did it mean to be "just a little plastic"? At the time, the consensus was: not very much. Plastics manufacturers had long known that additives could and would leach out of polymers but maintained that people weren't exposed to high enough levels to suffer any harm. After taking a hard look at DEHP and other phthalates, mainly in adults, independent toxicologists came to much the same conclusion. They found that very high doses could cause birth defects in rodents and induce liver cancer in rats and mice, but only through a mechanism that rarely affects humans. When I called Rubin and Jaeger, each told me that after much study they had concluded there was no cause for concern. Rubin, now retired, said he spent years studying DEHP and turned up only one relevant hazard: an uncommon phenomenon that surfaced during the Vietnam War in which injured soldiers who were in shock died after receiving blood transfusions from vinyl blood bags. Under that rare set of circumstances, the chemical could trigger a lethal immune reaction in the lungs. Otherwise, Rubin said, he'd come to the conclusion that DEHP and other phthalates "were — how shall I put it? — as harmless as chicken soup."

There were a few contrary voices at the time, but they mostly reflected a general sense of unease on the part of scientists who just weren't comfortable with the growing ubiquity of industrial chemicals and the intimate ways in which humans were coming into contact with them. It would take a sea change in the science of toxicology — and the late-life career change of a Colorado woman — before those voices would gain any kind of significant audience.

One of the basic textbooks of toxicology, which came out in 1987, clearly articulates the theory of poison that has held sway for centuries. On page 2 of *A Textbook of Modern Toxicology*, authors Ernest Hodgson and Patricia Levi declare that poison is "a quantitative concept. Almost any substance is harmful at some dose and, at the same time, is harmless at a very low dose." Hodgson and Levi use the example of aspirin: two tablets of aspirin are healing, twenty can cause an upset stomach, and sixty can be lethal. The dose makes the poison. This idea, first formulated by the medieval alchemist Paracelsus, has been one of the fundamental principles of modern toxicology.

But the very same year that Hodgson and Levi published the first edition of their toxicology text, a zoologist named Theo Colborn began developing a different theory of toxic effects that would challenge that conventional wisdom. Colborn wasn't even a toxicologist by trade. She had spent years in Colorado raising four children and working as a rancher and a pharmacist when she started to worry that pollutants in local rivers and lakes were contributing to health problems in the area. She wanted to better understand water-quality issues, so at the age of fifty-one, she went back to school and got a master's in freshwater ecology and then a PhD in zoology. In the late 1980s, she landed a job at the Conservation Foundation in Washington, D.C., where her boss asked her to survey research on the effects of pollution in the Great Lakes. It was a potentially mind-numbing job of data crunching. Yet Colborn, as one reporter observed in a profile, had a talent for mining the glittering patterns hidden in piles of data.

For months, she plowed through thousands of studies about the impacts of pesticides and synthetic chemicals on the Great Lakes wildlife. She expected to see off-the-chart rates of cancer, the classic telltale mark of a toxin. Instead, the reports were filled with weird, eerie accounts of chicks wasting away, cormorants born with missing eyes and crossed bills, male gulls with female cells in their testes, and female gulls nesting together. It seemed "like a hodgepodge of disconnected information," Colborn later recalled in a book she coauthored, *Our Stolen Future*. But she sensed "that something important was lurking beneath the confusing surface."

To try to make sense of the data, she created an electronic spreadsheet, sorting the information by species and health effect. Looking over the array of symptoms—reproductive failures, immune problems, abnormal behaviors—she saw a pattern. Most could be traced to a dysfunction of the endocrine system, the complex network of glands that produce the hormones (such as estrogen, testosterone, thyroid hormone, human growth hormone) that govern growth, development, metabolism, and reproduction. This was a new and potentially profound type of environmental hazard—especially in light of another pattern Colborn was seeing in the data: the adult animals exposed to chemical toxins were mostly fine; the main health problems were in their offspring. Unlike typical toxins, these seemed to be acting as what she called "hand-me-down poisons."

Such transgenerational events weren't unknown. As Colborn was well aware, the drug DES, a powerful synthetic estrogen, had had a similar effect. The drug had been widely prescribed to women in the 1940s and '50s to prevent miscarriages. But years later, the children who had been exposed to the drug in the womb began experiencing a host of health problems. DES daughters had higher than normal rates of breast cancer and an extremely rare vaginal cancer, as well as reduced fertility and other reproductive problems, while DES sons were prone to undescended testicles and hypospadias (when the opening of the penis occurs along the shaft rather than the head). The DES experience alerted scientists to the potential hazards of synthetic hormones.

But Colborn wasn't investigating a drug designed to mimic a hormone. Moreover, the pesticides and industrial chemicals she was examining had far wider circulation than DES. Her data suggested the frightening possibility that wildlife and people were being exposed to an entirely new kind of risk from widely used chemicals — one that challenged the Paracelsian paradigm that was the core tenet of regulatory testing. The poison wasn't solely in the dose; it could also be in the timing of exposure.

The implications of her hypothesis weighed heavily on Colborn's mind, and after finishing her report, she felt she couldn't just move on to another study. So she organized a meeting in July 1991 at the Wingspread Conference Center in Racine, Wisconsin, pulling together twenty top researchers from various fields whose work had shaped her assessment and who had the collective expertise to consider her theory. The group members came from a range of disciplines — biology, endocrinology, immunology, toxicology, psychiatry, ecology, and anthropology — that rarely talked to one another. "I was scared to death!" Colborn recalled. "There I was, a brand-new PhD who knew only a handful of wildlife biologists." But she pushed the group hard, working them from morning till night so they could get to know one another and make connections among their work. At the end of the weekend, the group agreed that, like the proverbial blind men feeling the elephant, each had been describing pieces of the same disturbing trend. They dubbed it "endocrine disruption." Its hallmarks included three important findings often overlooked by traditional toxicological research: the effects could be transgenerational; they depended on the timing of exposure; and they might become apparent only as the offspring developed.

The first Wingspread conference identified about thirty chemicals as endocrine disrupters. Today, the number may be anywhere from seventy to a thousand, depending on who is doing the counting. It's difficult to pin down because of the complex effects that result from interference with normal hormonal activity and the many mechanisms by which endocrine disrupters cause trouble. For instance, by mimicking natural hormones, they can insinuate themselves into special receptors on cells that activate certain genes. Or they can block a natural hormone en route to its destination, preventing it from delivering its chemical message to the cellular receptor. However many endocrine-disrupting chemicals there are, the ranks include several found in common plastics.

Aside from phthalates, one of the most infamous suspects at the moment is bisphenol A, the primary component of polycarbonate, a hard, clear plastic that's used to make a host of consumer items including baby bottles, compact discs, eyeglass lenses, and water bottles. Bisphenol A is also a basic ingredient of the epoxy resins used to line canned foods and drinks. Unfortunately, the bonds holding these long molecules together can be weakened fairly easily. Hot water and detergents can loosen the links in the polymer daisy chain, and when that happens, small amounts of bisphenol A can slip free. So each time a polycarbonate bottle is washed, a little bit of BPA is unloosed and is able to leach out. Scientists have known since the 1930s that bisphenol A acts as a weak estrogen, allowing it at least two possible ways to cause static in the body's normal hormonal conversations: by binding with estrogen receptors on cells and by blocking natural stronger estrogens from communicating with cells. Either can disrupt how the body uses and produces natural estrogen.

By now hundreds of studies have suggested bisphenol A does just that in animals and humans. Researchers have reported the compound causes health effects in cells and animals that are similar to diseases becoming more common in people, such as breast cancer, heart disease, type 2 diabetes, obesity, and neurobehavioral problems such as hyperactivity.

Bisphenol A research has been hugely controversial, in part because the purported effects seen at very low doses don't show up at higher doses — a complete contradiction of Paracelsus's famous dictum. Yet it makes sense if you view the chemical as a hormone rather than as a typical poison in which the toxic effects increase with the amount of exposure. So said Frederick vom Saal, a reproductive endocrinologist at the University of Missouri, who pioneered much of the research on the molecule. Hormones are produced in accordance with a finely tuned feedback system that's regulated by a pair of command-and-control glands in the brain, the pituitary and the hypothalamus. If the levels of a hormone get too high or too low, the hypothalamus relays that information to the pituitary, which in turn signals the gland that produces that hormone to gear up or slow down. Because of that feedback loop, said vom Saal, "sex hormones cause opposite effects at high and low doses. That's what we teach undergraduates. At high doses, they turn off responses that they stimulate at low doses."

In the case of bisphenol A and a few other plastics, the worrisome element is an integral part of the molecular structure. Likewise, some critics consider polystyrene potentially dangerous because it is built from a monomer, styrene, that is a known neurotoxin and a suspected carcinogen, and a few studies have suggested styrene is also capable of migrating out of the polymer chain.

But most of the compounds that have experts worried are additives, chemicals such as phthalates that are blended into base polymers to confer desired properties. If they are only lightly linked to the daisy chains, they are apt to escape from the polymers. The phthalates in PVC, for instance, are drawn to lipids, or fats, and will fairly fly off the polymer chain so they can dissolve in those fatty molecules. Manufacturers rely on a vast assortment of fillers, fluffers, antioxidants, dyes, fire retardants, lubricants, stabilizers, plasticizers, and other additives to fine-tune their plastic products — so many that one reference book on polymer additives runs 656 pages long. So many that plastics additives themselves constitute a nearly thirty-seven-billion-dollar global market.

A few of the chemicals used in plastics — the phthalates found in IV bags; triclosan, an antibacterial found in kitchenware and toys; and the brominated fire retardants widely used in furniture — have already caught the attention of researchers and regulators. But that leaves hundreds, if not thousands, that we know little or nothing about. Witness the recent experience of German researchers who, much to their surprise, found that water bottles made of polyethylene terephthalate (PET) seemed to leach small amounts of one or more unknown compounds that mimic estrogen. "We knew there are plastics that release endocrine disrupters, but we didn't expect [to see that] in PET," study author Martin Wagner told me. He didn't try to identify the compound causing the effect, but one possibility is antimony, a chemical catalyst used to make PET that has been shown to have estrogenic activity.

Unfortunately, there's no way for consumers to know what chemicals are in the plastic things we buy. Manufacturers generally aren't required to list the ingredients of their plastic products. Indeed, given the long supply chain from raw polymer to finished product, most likely they don't know what's in the plastic resins they've used. (The resin codes on packaging were designed to aid recycling; they offer limited information, at best.) For the most part, consumers are as clueless as incubator babies about the plastics that surround us, a point that was driven home to me one day when I got a new plastic floor mat from OfficeMax.

It had a faintly chemical odor at the store, but after I left it in the car for a couple of hours while I ran other errands, the smell was overpowering. I pulled out the box and turned it all around, vainly hoping for some indication of what was in the mat. The box said only that it contained a blue chair mat with the helpful added note *chair not included*. It was a good bet the mat was vinyl, but I doubted the smell could be caused by any of the phthalates used in vinyl because, although they off-gas, they are odorless. If you sniff an IV bag, you won't smell anything. Was the smell something I should be worried about or was it merely annoying? I had no way of knowing. When I later called OfficeMax, the company confirmed the mat was made of vinyl but could offer no more information. Allen Blakey, a spokesman for the Vinyl Institute, suggested the scent was due to inks and sulfides added to the plastic. "Some people love that smell," he said, adding that the odor dissipates after a few days.

I try not to be alarmist about these things, recognizing that my life is rife with more tangible risks. I grew up in a household of smokers and smoked for many years myself. I sometimes talk on my cell phone while driving. My house is prone to mold. I forget to put on sunscreen. After my trip to OfficeMax, I was going to get onto a freeway swimming in diesel particulates and exhaust, not to mention teeming with speeding cars. I considered dumping the floor mat, but then I'd be out thirty-nine dollars. In the end, I decided just to roll down the windows and wait for the wind to carry the smell away.

Many suspected endocrine disrupters interfere with estrogen. However, DEHP, the chemical found in IV bags and tubing, is an antiandrogen, meaning it interferes with testosterone and other masculinizing hormones coursing through the bodies of both men and women. Medical devices may be a significant source of exposure, but most of us come into contact with DEHP through its nonmedical deployment in such vinyl items as shower curtains, wallpaper, venetian blinds, floor tiles, upholstery, garden hoses, swimming-pool liners, rainwear, car upholstery and convertible tops, and the sheathing on cables and wires. It's been found in flip-flops and plastic shoes, modeling clay such as Fimo and Sculpey, yoga mats, cosmetics and nail polish, cleaning products, lubricants, and waxes, not to mention household dust. But our primary exposure is through fatty foods, such as cheese and oils, which are particularly likely to absorb the chemical, though it is unclear whether that is happening via plastic packaging, the inks used in food wrapping, or during commercial preparation and processing. For instance, DEHP in milk has been traced to the tubing used by dairies.

Such ubiquity means the chemical can get into our systems through almost any route — by inhalation, ingestion, or absorption through the skin. Once the compound enters the bloodstream, it is broken down into smaller molecules, called metabolites. These metabolites are actually the toxic troublemakers. They're small enough to be absorbed by cells, including, most significantly, cells in the pituitary gland. The pituitary is the Leonard Bernstein of the endocrine system, the gland that conducts the complex symphony of hormonal releases by other glands and cells. DEHP takes a seat and promptly behaves like a wayward violin, introducing discordant sounds that clearly don't belong. Among other things, its metabolites stop the pituitary from producing a hormone that directs the testicles to make testosterone. When that occurs during sensitive periods of development, testosterone levels throughout the body can plummet, which in turn can trigger an avalanche of effects — at least in developing animals.

For example, researchers at the Environmental Protection Agency administered DEHP and another common phthalate, DBP, to male rats in utero during the period when sexual differentiation occurs; previous studies had focused on other phases of development. The scientists found that the chemicals caused dramatic changes in the fetuses' reproductive systems. The pups were more likely to be born with undescended or missing testicles, low testosterone levels, and reduced sperm counts. They also were prone to a shortened distance between the anus and the penis, hypospadias (abnormal opening in the shaft of the penis), and other malformations. The cluster of symptoms was striking enough that the researchers gave it a name: the phthalate syndrome. In further studies, they found that afflicted rats were more likely to suffer impaired fertility and develop testicular tumors later in their lives. Though the effects are most pronounced in males, researchers found that female rat pups could also suffer from DEHP exposure, developing cysts on their ovaries and ceasing ovulation.

As Colborn would have predicted, the devil was in the timing of exposure. The EPA studies used relatively high doses of phthalates, but other animal studies have found that even very small amounts of DEHP can reduce sperm production, provoke early puberty, and have other subtle warping effects if administered during critical periods of development. That's when being a little plastic can be a big problem.

The effects in rats are mirrored by broad trends in human reproductive health. Epidemiologists have charted rising rates of male infertility, testicular cancer, decreased testosterone levels, and diminished sperm quality in many Western countries. The number of boys born with hypospadias has increased significantly since the late 1960s, according to some studies (though not all), and the number born with undescended testicles has also risen during the same period. Meanwhile, several studies have reported that infertility rates among women are on the rise. During this same period, humans' exposure to DEHP and other phthalates has undoubtedly increased, given their steadily rising production and use. The result is that at least 80 percent of Americans — of all ages and races, from city dwellers to residents of remote rural towns — now carry measurable traces of DEHP and other phthalates in their bodies, according to biomonitoring studies by the Centers for Disease Control. Researchers have detected phthalates in blood, urine, saliva, breast milk, and amniotic fluid, which means people are being exposed to the chemicals at all stages of life, starting in utero. The chemicals pass quickly through our bodies, but we are exposed on such an ongoing basis that the overall amounts in our systems don't change dramatically over time. Our cells are under continual low-level chemical assault.

None of us are exposed to DEHP at the levels that produced those devastating results in fetal rats. Still, many of us are taking in more than the daily limit advised by the EPA — a threshold set in 1986 and based on studies done well before the concept of endocrine disruption was on anybody's radar. Some of the people with the highest levels are the very ones who most need to steer clear of hormonal disrupters, such as children and women of childbearing age. Studies consistently find that children harbor higher phthalate levels than adults, which may be due to several factors, including that they metabolize chemicals more slowly; eat, drink, and breathe more per pound of body weight; and are more likely to put vinyl toys in their mouths. (One researcher told me she once found a particularly noxious-smelling vinyl bath book for babies called *Splish-Splash Jesus*.) Such findings led an expert panel convened by the National Toxicology Program in 2006 to conclude that there were grounds for "concern" that DEHP exposure can affect the reproductive development of baby boys under the age of one.

But the group that seems to be at greatest risk are newborns undergoing treatment in a NICU. Studies show that a baby like Amy who is hooked up to IV bags and tubing for weeks at a time may end up absorbing doses of DEHP a hundred to a thousand times higher than the general population. The chemical hit will be even larger if she gets blood transfusions or if she later has to be connected to a heart-lung machine that circulates and oxygenates the blood in her body. The machines are commonly used for babies with severe respiratory problems or those who have undergone heart surgery. They're lifesavers, but they also deliver huge amounts of DEHP, said Naomi Luban, a pediatric hematologist at Children's National Medical Center who works with Short and has long been concerned about the risks of plasticizers. The transfused blood is laced with DEHP from blood bags, but there's another problem as well. As the blood circulates through the heart-lung device, it flows through several dozen feet of plastic tubing and a plastic membrane — all leaching DEHP. A really sick baby getting that kind of intensive treatment can wind up with a cumulative exposure twenty times higher than the level deemed safe for human intake. (Indeed, people of any age undergoing procedures such as kidney dialysis and blood transfusions are at risk of getting a significant blast of DEHP.)

Because these newborns are underdeveloped, they may be especially vulnerable to the chemical's hormonal impacts. The cellular barriers to their brains and organs are more easily permeated. What's more, they don't yet have the capacity of adults or even of older children to clear the chemical from their systems, meaning it circulates for a longer time, increasing the potential for damage.

As the CDC has pointed out, the mere presence of DEHP, or any chemical, in someone's blood or urine does not mean it is a health hazard. The big, and difficult, question is whether the small amounts to which we are all regularly exposed are sufficient to affect *some* people's health. We may all be a little plastic, but that doesn't mean we are all affected in the same way. Some individuals, such as NICU babies, may be at greater risk because of the amounts they take in or their stage of life. Given that researchers can't subject humans to the kinds of tests done with lab rats to pinpoint the conditions that cause adverse effects, it's not unreasonable to view ourselves as subjects of a vast, uncontrolled experiment. But we are not entirely blind as we journey through this vast plastic laboratory called modern life. Epidemiological studies — surveys of large groups of people — offer an indirect way to gather evidence. Shanna Swan, a reproductive epidemiologist at the University of Rochester School of Medicine and Dentistry, has conducted several such studies, and her findings suggest that some of us do pay a price for being a little plastic.

In one study, she measured phthalate levels in 134 pregnant women and then later closely examined the genitalia of their baby sons. She found the sons of mothers with the highest phthalate levels had subtle but unmistakable symptoms echoing the phthalate syndrome seen in rats. They were more likely to have undescended testicles, a smaller penis, and a shorter distance between the base of the penis and the anus, the measurement that in the rat studies was considered a hallmark of decreased fetal testosterone levels. "These babies had no abnormalities that a clinician would recognize," Swan emphasized. But given the long-term effects on exposed rats, she considered even these subtle changes a worrisome collection of symptoms that could affect the boys' fertility later in life.

Swan then decided to look at how else an antiandrogen might

affect a developing fetus. The male reproductive tract isn't the only body system touched by testosterone. Like estrogen, the hormone circulates throughout the body, affecting metabolism, growth, behavior, and cognition, as well as the actions of other hormones, in both boys and girls. "The brain is the largest sex organ," Swan likes to say. "It also is developed under the influence of testosterone."

Normally, testosterone levels in the brain surge during certain critical points of development, which is thought to play an essential role in the process of sexual differentiation. Studies have shown that when pregnant rats are exposed to drugs known to block that hormonal surge, their male offspring don't engage in the same kind of rough-and-tumble play as unexposed pups.

Taking her cue from those findings, Swan went back to the same group of parents and children she'd studied before. The kids were now in preschool. For this study, she had the parents fill out a detailed questionnaire about how their children play. She asked them to rate how frequently their child played with toys like dolls or trucks, how often the child set up house or pretended to fight. The boys with the highest fetal exposure to the phthalates DEHP and DBP had the lowest scores on typical boy play, such as pretending to shoot a gun. They were also more likely to prefer gender-neutral play, such as working on puzzles. The girls showed no effect.

It was the kind of finding headline writers love. "Common Chemicals Make Boys Soft," one Australian paper declared. But Swan wasn't reporting a nifty solution to macho aggression. She was describing a subtle shift in the hard-wiring of the boys' brains so that they played in "less typically masculine" ways. It was a small effect but one with potentially profound implications, given that testosterone, estrogen, and other hormones shape the many differences in how male and female brains develop and process the world.

Both studies need to be replicated. But having found that phthalates can affect two very different body systems, Swan asked, "Why should we assume the effects are just limited to two? My concern is we are seeing changes throughout the body wherever testosterone matters, and that's many, many places."

Other epidemiological findings lend weight to her concerns. A variety of small studies have found an association between exposure to phthalates and obesity, early puberty, allergies, ADD, and altered thyroid function — all conditions that could plausibly be related to hormonal disruptions. Most of the studies have focused on boys. But a handful suggest that girls may also be affected by the drops in testosterone or by some as-yet-undescribed effect on their estrogen levels. Some researchers have postulated that the chemical can suppress estrogen production in females, said Swan. "We're all struggling to figure out what's going on with females. It's difficult because the female reproductive system is invisible. The males are easier because it all hangs out." Still, a small number of epidemiological studies have found correlations between phthalate levels and endometriosis, miscarriage, uterine fibroids, and premature breast development.

DEHP's hormonal effects may not be the only problem. A 2010 study suggested that in very young infants, the chemical may interfere with the cellular systems involved in controlling inflammation, part of the body's method of fighting infection. Other studies have linked DEHP with immune system and respiratory problems and have continued to raise red flags about toxicity to the liver, especially in premature babies who are being exposed due to their treatment in NICUs. German researchers showed that babies in intensive care who received fluids from IV bags containing DEHP were more likely to develop a particular type of liver problem than those whose IV bags didn't contain the chemical.

This all sounds like pretty strong evidence, right? And yet science rarely delivers straight slam dunks. Consider just one endpoint thought to be affected by DEHP: cells in the testes that secrete testosterone. Rat studies have repeatedly found that DEHP damages those cells. But rats are the most phthalate-sensitive of all species tested. Recent primate studies involving young marmosets have found no such effect. Does that mean primates — our closest relatives — aren't as sensitive to the chemical as rodents? Or were the marmosets past the age when they would have been vulnerable to the effects? Researchers are still debating the question. Likewise, the epidemiological findings on sperm quality have been inconsistent: some studies show correlations with phthalate levels, some don't.

It's hard and expensive to mount the kinds of studies that can deliver clear-cut answers. For instance, Drs. Short and Luban have long wanted to do a study following babies who were heavily exposed to DEHP as a result of being on heart-lung machines. "If any population would have long-term reproductive effects, it would be those kids," said Short. "If the results came back negative, it would put all this stuff to rest." They did a small pilot study in which they found and tested eighteen teenagers who as infants had spent time in the NICU on the machines. None showed signs of any reproductive problems. But it's impossible to draw any valid conclusions from a group that small. Statistically speaking, at least two hundred and fifty kids would have to be studied to get robust results. Short and Luban estimated it would take ten million dollars to track down and test that many survivors. They wrote up a proposal to do such a study, but neither the National Institutes of Health nor private industry was willing to make the investment. "That ten-million-dollar price tag was just prohibitive," said Short.

"But, boy, it would have been nice to have [the answers]," added Luban.

The continuing uncertainties are one reason why expert panels that have looked at DEHP and other phthalates have come to differing conclusions and also why nearly every research paper ends with the same mantra: more and better studies are needed.

One of the biggest gaps in research is the paucity of studies looking at the real world of chemical exposure. A person is not exposed to just a single chemical at a time; each day each of us encounters hundreds. And that chemical bombardment begins even before birth: a study by the organization Environmental Working Group found an average of two hundred industrial chemicals and pollutants in the umbilical-cord blood of ten newborns. What's the cumulative effect? Researchers have only begun addressing that question. The early findings are cause for concern.

Earl Gray, the EPA researcher who identified phthalate syndrome, tested mixtures of phthalates in rats. He deliberately used low doses of each, well below amounts that could produce effects individually. Yet when he exposed male rats in utero to the mixtures, as many as 50 percent were born with hypospadias or other reproductive abnormalities. In combination, the chemicals were far more potent than they were individually, suggesting that compounds that act on the same hormonal pathways have an additive effect, he said.

Researchers say we need more of these kinds of studies, ones designed to mimic the real experience of human exposure. Swan and others want to see more research that focuses on pregnant women and children to gain a long-term picture of chemical effects rather than isolated snapshots. That's just what a recently begun study aims to do: the National Children's Study will follow a hundred thousand children across the United States from birth to the age of twenty-one in an effort to tease out environmental influences, including exposure to phthalates and bisphenol A, on health.

So if DEHP and other phthalates haven't been proven unsafe, does that mean they are safe?

The chemical industry, as might be expected, maintains they are. There is, after all, a $1.4 billion market in phthalates on the line. The American Chemistry Council's position, as one spokeswoman noted, is that "DEHP medical devices have been used for better than fifty years, and there hasn't been any verified evidence of harm to humans." Even in the case of neonates, the group maintains, the benefits of treatment outweigh the risks posed by exposure.

The ACC vigilantly tracks the research, publicizing studies that show no adverse effects and picking apart each finding that suggests a problem. It criticized Swan for using "unproven methods," such as the anogenital distance measurement and the play survey, and hammered on methodological errors, which she acknowledges but insists did not affect the statistical significance of her final results. In general, the ACC draws on a standing set of criticisms to point out flaws that may be accurate but not always meaningful. Among the oft-cited complaints: the sample sizes are too small; rats are poor models for human health hazards; the doses administered in animal studies are much higher than those experienced by humans; the demonstrated health effects are not necessarily adverse. Almost invariably, when an epidemiological study reports a risk associated with phthalates, the group counters with a press release pointing out that the study shows a correlation only, not proof of a causal effect. Which is true — that's precisely what epidemiological studies do. Still, the correlations highlighted by epidemiological studies have long been the gold standard for assessing risks to public health.

The insistent focus on the flaws of each individual study ignores — and obscures — how each may be contributing to an increasingly disturbing body of evidence. The nitpicking zeroes in on and tries to amp up the uncertainty that is always inherent in science. It's a strategy taken straight from the tobacco industry, and one that, incredibly enough, was committed to paper in 1969 by an executive at the cigarette maker Brown and Williamson: "Doubt is our product since it is the best means of competing with the 'body of fact' that exists in the mind of the general public."

As Swan and others have pointed out, there was never a single study that "proved" smoking causes lung cancer. It took a combination of in vitro, animal, and human population studies to demonstrate the dangers of tobacco. A series of epidemiological studies pointed to the risk, prompting the surgeon general to issue his famous 1964 warning. Over the next forty-some years, researchers painstakingly pieced together, through cell and animal studies, the biological mechanisms that explained how tobacco smoke could induce tumors in the lungs. Meanwhile, the tobacco industry spent those decades denying there was any connection.

It's unlikely science will deliver neat, definitive answers about the risks of endocrine disrupters any time soon. So do we wait for children like Amy to grow up before we discover the dangers of DEHP exposure? Do we wait to see if she develops liver problems, launches into puberty at an early age, has trouble conceiving a child? Or have we reached the point where there is enough evidence to fairly act on the side of caution? I think we have. Yet our current system for regulating chemicals makes that difficult to do.

We have no coherent or comprehensive body of law for managing the chemicals we experience in daily life. Instead, there's a weak and uncoordinated patchwork of national and state laws. Federal regulation of chemicals is divided among various agencies, leading to fragmented and inconsistent policies. The EPA, for instance, recently announced it would take steps to limit use of phthalates, including DEHP. The FDA, however, still judges that the chemical offers more benefit than risk and has thus far ignored calls to limit its use in medical devices and require labeling of medical products that contain it. The FDA's only action to date has been a 2002 advisory recommending that hospitals not use devices containing DEHP in women pregnant with boys, in male infants, and in young teenage boys. U.S. regulators at both the FDA and the EPA have lagged behind the changing scientific understanding about chemical risks. For instance, both agencies still base their safety assessments of chemicals on studies of one chemical at a time, rather than looking at aggregate effects.

But there is a much bigger problem: U.S. law tends to treat chemicals as safe until proven otherwise. Regulators are required to find what author Mark Schapiro called "a scientifically improbable smoking gun" before they can pull a suspect chemical off the market. Nowhere are the failings of that approach more clear than in the main federal law regulating synthetic chemicals, the Toxic Substance and Control Act. Enacted in 1976, the law gives the EPA the power to require testing of and restrict chemical substances. Yet it's had little opportunity to exercise that power. The sixty-two thousand chemicals in use when the law was passed were exempted from the testing requirements. And the law's provisions bind EPA regulators in a Catch-22: they need evidence of harm or exposure before they can require a manufacturer to provide more information about a chemical, but without the information, how do they establish evidence of harm? In the absence of that evidence, the regulators cannot act. So while twenty thousand chemicals have been introduced since 1976, the EPA has been able to require intensive reviews for only two hundred, and it has used its authority to restrict only five. The hurdles are so high, the agency could not even successfully ban asbestos, an undisputed carcinogen. "This means," wrote John Wargo, an expert on chemical policy, "that nearly all chemicals in commerce have been poorly tested to determine their environmental behavior or effects on human health."

Are *all* those chemicals dangerous? It's hard to tell, but the EPA has said that at least sixteen thousand were potentially causes for concern owing to their high production volume and chemical properties. Meanwhile, European regulators have estimated that an astounding 70 percent of new chemicals have some hazardous property, ranging from carcinogenicity to flammability.

Everyone concerned with chemical policy — the head of the EPA, environmental activists, even the American Chemistry Council — agrees the law is a poor vehicle for navigating our current chemical landscape. Yet agreeing on some alternative approach is another matter (as has been evident in debate over a reform bill winding its way through Congress as of mid-2010). A major point of contention is that American policymakers are starting to look to Europe as a model for regulating the chemical industry.

In Europe, the burden of proof is on safety rather than danger. European regulators "act on the principle of preventing harm before it happens, even in the face of scientific uncertainty." Guided by that precautionary principle, Europeans began limiting DEHP and other phthalates while American regulators continued debating the risks. (The EU, for instance, barred the use of DEHP in children's toys in 1999, nine years before the U.S. Congress passed similar legislation.) A new directive known as REACH (for Registration, Evaluation, and Authorization of Chemicals), adopted in 2007, requires testing of both newly introduced chemicals and those already in use, with the burden on manufacturers to demonstrate that they can be used safely. The agency charged with implementing REACH targeted DEHP as one of the first fifteen "substances of very high concern" to be regulated. In essence, European regulators are treating chemicals the way U.S. regulators treat drugs: they're presumed to be dangerous unless shown to be otherwise. American manufacturers are already selling products in European markets that have been reformulated to comply with the precautionary principle. Shouldn't American citizens demand the same on this side of the Atlantic?

Some states have already taken the initiative. In 2008, California passed landmark safer chemicals legislation that requires the state to collect data on chemical toxicity, restrict some of the most hazardous substances, and promote research into safer substitutes. The law expands on the approach taken by Massachusetts, which since 1989 has required companies using large quantities of toxic materials to disclose that use and explore alternatives to dangerous chemicals and has promoted programs to help companies either switch to safer alternatives or use smaller quantities of the hazardous substances.

Still, state efforts are no substitute for across-the-board federal protections. John Wargo, a professor at Yale University, argued convincingly that what we need is a national "plastics control law." Congress, he pointed out, has passed laws to regulate other health or environmental risks, such as pesticides, pharmaceuticals, and tobacco. Why not the same for plastics — materials that touch the life of every single American? He proposed a comprehensive policy that, among other things, would include tough premarket testing of chemicals used in plastics, mandatory labeling of ingredients, and strict prohibitions of chemicals and compounds that pose a threat to human health or that don't quickly degrade into harmless substances.

Of course, it's possible to set the safety bar too high as well as too low. Plastics have proven benefits in medicine and other fields. Proposing unachievable standards — demanding that substances be proven absolutely benign — could produce the same regulatory "paralysis by analysis" we're now experiencing. But when adequately funded research indicates significant evidence of harm, we are betraying our children if we fail to take action against a potentially dangerous chemical, especially when alternatives are available.

Who's responsible for taking action? Often the first step is made by an ordinary person paying attention. A person like Paula Safreed.

In the early 1990s, Safreed was a nurse in the NICU at the Brigham and Women's Hospital in Boston. In addition to caring for babies, she was in charge of ordering supplies for the unit, which put her in regular conversation with the vendors of medical equipment. She began hearing rumblings about the chemical in the plastic IV bags and tubing she used to treat her tiny patients. What caught her attention wasn't anything to do with DEHP's potential reproductive effects. Rather, it was a comment by a vendor that DEHP had been linked with damage to the liver. It reminded her of a baby she had cared for years earlier, one of a set of twins born prematurely. One twin died, but this boy survived and spent months in the Brigham NICU, receiving blood transfusions, proteins, and lipids, as well as intravenous nutrients, via vinyl IV bags and tubing. Safreed was thrilled to see him get strong enough to go home and utterly devastated when he later developed liver cancer and died at the age of three. She knew he had liver troubles, a common complication in preemies who spend a long time taking artificial nutrients. Still, Safreed was haunted by the possibility that the very things the NICU staff did to save his life had contributed to ending it.

"That was when I started to get interested in plastics," she said. She began quizzing vendors about the contents of their products and pushing the hospital administrators to buy bags and tubing and other equipment made without DEHP. At the time, alternatives were few and far between and significantly more expensive. Still, Safreed kept pushing.

She was not alone. During that same period, a coalition of environmental groups started a new organization, Health Care Without Harm, aimed at getting hospitals to phase out their use of vinyl IV bags, tubing, and other equipment that contained the plasticizing chemical DEHP. The drive was an outgrowth of a broader campaign to get hospitals to stop using PVC in general, because incineration of medical waste containing PVC had made hospitals a leading source of dioxin emissions. Discovery of that fact was "an incredible irony and a teachable moment," according to Health Care Without Harm founder Gary Cohen. "Because if you want to detoxify the economy, you start with the sector of people who have taken an oath to do no harm."

Organizers from the group began talking about the risks of DEHP with hospitals around the country, quickly winning the support of influential university-affiliated hospitals such as the Brigham, as well as powerful chains like Kaiser Permanente and Catholic Healthcare West, both of which made far-reaching pledges to completely eliminate PVC and its additive DEHP from their facilities. (Kaiser has even replaced its vinyl carpet and flooring with alternative materials.) As of 2010, about 120 of the more than 5,000 hospitals in the United States had publicly signed on to Health Care Without Harm's campaign.

The Brigham NICU was one of the nation's first to start adopting alternatives to PVC, and nurses there give the credit to Safreed, who has since retired. "All our products are DEHP-free," the NICU's assistant manager Julianne Mazzawi said proudly while showing me a package containing a slender coil of IV tubing that had a prominent label to that effect. The FDA may not require labeling yet, but some manufacturers are going ahead and doing it on their own.

But what's really driven change in the medical marketplace has been Health Care Without Harm's success in reaching the half dozen or so organizations that negotiate bulk purchases on behalf of most of the nation's hospitals. These group-purchasing organizations hold enormous sway over the market; when they began asking about alternatives to PVC and DEHP, medical manufacturers sat up and listened.

Most major medical suppliers now offer products, particularly many types of IV bags, tubing, and neonatal care equipment, that are free of DEHP and PVC. Some suppliers, such as Baxter, have introduced alternatives while continuing to defend the safety of PVC and DEHP, which lends a certain schizophrenic quality to their promotional brochures. Other suppliers embraced new approaches early and with gusto. Starting in the 1970s, leaders of the company B. Braun (now B. Braun McGaw) saw the opportunity to carve out a new market niche by developing alternatives to PVC and DEHP.

It mainly uses the common plastic polypropylene (the stuff of bottle caps, disposable diapers, and monobloc chairs). "Polypropylene is a cleaner material than vinyl because it doesn't contain chloride. And it doesn't contain plasticizers" so there's nothing to leach out, said David Schuck, vice president of pharmaceutical testing for the company. He said safety testing has shown the resin has no hormonal effects. (Still the company uses glass containers only for its intravenous nutritional supplements for infants.) Other companies are using other types of plastics — such as polyurethane, polyethylene-like polymers, and silicone — that they say are safer and don't require the use of chemical additives.

Meanwhile, the makers of additives are coming out with alternative, ostensibly safer plasticizers that can be used to soften PVC. At least four are already used in children's products, including citrates, compounds based on citric acid. These have been available for years but were rarely used because they cost more than phthalates. Another alternative softener is a chemical called Hexamoll DINCH, introduced by BASF, one of the world's leading makers of phthalates. BASF spokesman Patrick Harmon said the company spent seven million dollars on safety testing and is "very confident" it is safe. Though American manufacturers have not yet adopted it, he said it is being used in Europe for toys, food packaging, and medical devices, including the IV tubing and bags used to deliver nutritional supplements to preemies.

Given the growing availability of alternative materials and additives, I was surprised to find they constitute only about a quarter of the medical market. There are many hospitals — and even neonatal intensive care units — that are still using equipment that contains PVC and DEHP. They are understandably cautious about trying something new — better to stay with the devil you know. And it's a market reality that alternatives still cost more. In the absence of a federal mandate — the FDA warning was only an advisory — it's not hard to see why cash-strapped hospitals would be reluctant to make a change.

The NICU at Children's National has switched where it can, said

Short. The babies now get nutritional supplements from non-PVC bags — an important changeover, because those fatty liquids are particularly good at sucking DEHP out of vinyl. But Short said she hasn't found a satisfactory alternative for the tubing used with those bags or for the regular kits of IV bags and tubing that were essential to baby Amy's care. "The tubing has to be soft," she explained, rubbing her fingers along one of the impossibly thin pipelines running into Amy's impossibly tiny veins.

And there are also applications for which no alternatives have yet been found. Heart-lung machines still use only vinyl tubing that contains DEHP. Ironically, the same is true for the object that first triggered critics' scrutiny: the blood bag. DEHP, it turns out, acts as a preservative for red blood cells, preventing them from breaking down. To many blood bankers, the fact that DEHP leaches out of the bags is a plus, not a minus, said Gary Moroff, a spokesman for the American Red Cross. No other plasticizer works as well, he insisted.

"That's because no one has evaluated them!" Luban responded in a tone of plain aggravation when I told her what Moroff had said. She has long been frustrated with the blood bankers' stay-put position and their apparent lack of interest in further investigating the potential risks of DEHP. But she also acknowledged that changing the basic material used in blood banking would require "an astronomical amount of work, time, effort, and money." The alarms aren't yet ringing loud enough to motivate anyone in the blood-banking field to undertake such an immense project.

Are the alternatives to phthalates safer than the chemicals they're replacing? One would hope so. DINCH has passed the scrutiny of EU regulators, and citrates seem to have a safe track record. But absent reliable methods of assessing chemical risks and without a precautionary chemical policy, there's no way to be certain that the substitute introduced today won't turn out to be the DEHP of tomorrow.

Markets are an unreliable force for protecting the broad public interest. They respond to public pressure, but the public is not always equipped to provide it. Until Americans have stringent laws in place designed to prevent harm before it occurs, we will be left to navigate a world of imperfect choices. For instance, even those neonatal intensive care units that have largely eliminated DEHP and PVC may still harbor other plastic-based risks. In a recent study of the vinyl-free Brigham NICU, researchers found the babies had bisphenol A in their urine. The source is unclear, according to Steve Ringer, the chief of the unit and one of the authors of the study. But given the ubiquity of bisphenol A, there are many possibilities, including the polycarbonate incubators, the plastic containers that hold the fats the babies are fed, and even the babies' mothers' breast milk.

If breast milk turns out to be a significant source, what should the hospital do?

"Our opportunities for intervention are pretty limited," Ringer said. The babies could be given formula instead, but in light of the uncertain, long-term risks posed by bisphenol A and the well-established, immediate health advantages of breast milk, he said, "I think I'd still come down on the benefits of breast milk."

And of course, even if market pressures could succeed in completely eliminating suspect chemicals like bisphenol A and DEHP from medical settings, we're still going to encounter them in every corner of our daily lives. It's nearly impossible to escape the plastic bubble.

When I consider how much of our lives are a little plastic, I find myself confused about how to negotiate this new world of risks. That's one reason I asked most of the researchers I interviewed how they dealt with plastics.

Knowing that there are bloggers and websites urging wholesale abandonment of plastics, I was surprised at how moderate most of the experts were. Joel Tickner, an associate professor of environmental health at the University of Massachusetts in Lowell and a longtime critic of DEHP, was typical. "I try to be as careful as I can, but I'm not obsessive," he explained. When his kids were little, he let them have vinyl squeeze toys—the sort that typically contain phthalates—but not a lot of them. He uses Tupperware, plastic wrap, and baggies. He doesn't heat or microwave foods in plastics, since that accelerates the breakdown of the polymer. (Every expert took that precaution, at a minimum.) Tickner swapped out his children's bisphenol A–containing polycarbonate water bottles for metal ones, but he isn't too worried about using them himself. He's careful about the amount of canned foods his family eats — to avoid bisphenol A — but hasn't banished cans from the pantry. When his son had hernia surgery, he checked to see if the IVs and other equipment contained DEHP. As it turned out, they didn't, but even if they had contained phthalates, Tickner would have been sanguine. It was a one-time procedure, not ongoing therapy, and the benefits clearly outweighed the risks. "You do the best you can do, while keeping in mind there is no easy answer," Tickner said. "I'd rather change the rules so people don't have to worry about this than spend all my time worrying about it."

Matter Out of Place

L IKE SO MANY CHILDREN, when I was young I was captivated by the idea of a message in a bottle. There was something appealingly random yet intimate about this most primitive form of long-distance communication. The thought that you could launch a missive from an American beach and that someone might pick it up in China or Tanzania or Ireland made the vast world suddenly seem much smaller and more negotiable. The oceans connected us all.

I was reminded of that point recently when I was talking to a researcher about a remote stretch of the Pacific Ocean northeast of Hawaii. For a long time, this area was known only as a windless becalming spot that sailors tried to avoid. More recently it has gained prominence as the site of an enormous swirl of plastic trash. The researcher and his colleagues had gone there to investigate just how bad conditions were in the so-called garbage patch. Twice a day they trawled nets through the clear blue waters, and every trawl brought up plastic. Most of what they recovered was unidentifiable bits and pieces, but one day, he said, the net contained a clear plastic disposable lighter. It was in such good shape you could even see an address printed on its side. He told me there was a picture of it on his lab's website.

I looked up the photo and peered closely at the writing—which was in both Chinese and English—until I finally made out a Hong Kong address and phone number. Feeling as if I'd found a message in a bottle, I decided to give the number a call.

It turned out to belong to a company that sells and distributes Chinese wine. Alex Yueh, the very patient office manager who surely regretted taking my call, didn't know about any promotional lighters. The company didn't give them out. But maybe it did in the past, he suggested, for the address listed on the lighter was where the company had been located seven years ago. "I don't know how this got into the ocean," he said. "It is very strange."

But in fact, in the age of plastic, the strange becomes commonplace: a lighter designed for a lifetime of mere months will easily survive for years bobbing across miles of open ocean.

One place that a castaway lighter could land is on Midway atoll, a tiny island among many in the Hawaiian archipelago. Midway is the site of a historic World War II battle. Today it is under a very different kind of assault from the piles of trash that wash up there with every storm. Volunteers collect tons of debris from its white-sand beaches each year. They have gathered hundreds of beached lighters.

Midway is also home to the Laysan albatross, an impressive sea bird that stands nearly three feet tall and has a wingspan twice that length. Those broad wings allow the birds to fly great distances in their daily forage for food in the seawater surrounding Midway. Some 1.2 million of the birds nest on the atoll, and almost every one has some quantity of plastic in its belly. The contents of the birds' stomachs "could stock the check-out counter at a convenience store," said one expert.

John Klavitter, a wildlife biologist with the U.S. Fish and Wildlife Service who has worked on Midway since 2002, has autopsied hundreds of dead albatrosses. He routinely finds bottle caps, pen tops, toys, fishing line, plastic tubes used in oyster farms, and lighters—all scooped up accidentally by the birds in their hunt for the squid and clumps of flying-fish roe that make up their normal diet. "And then there are all sorts of little plastic pieces, you can't tell what they are," Klavitter added. While rooting through the bird's stomach in the course of an autopsy there's often a sickening sound of clinking plastic. One dead chick had more than five hundred pieces of plastic in its stomach, including an olive green tag that was traced to a U.S. Navy bomber shot down more than ninety-six hundred kilometers away — in 1944!

A century ago, the biggest danger to the birds was feather hunters. Today, a major threat is plastic. The most vulnerable are the chicks, who rely on their parents for nourishment. Normally the parents regurgitate slurries of squid and fish eggs scavenged from the open sea into the mouths of their young. But since scientists started checking, in the 1960s, the birds have been returning to the nests with growing amounts of plastic. "If you sit and watch the parents feeding the chicks, you can see them passing plastics to their chicks," said Klavitter. In one two-month cleanup of the areas around the rookeries, volunteers collected more than a thousand disposable lighters.

Death is common among albatross chicks. Of the five hundred thousand chicks born each year, about two hundred thousand die — most often of dehydration or starvation. But a new factor likely contributing to that death toll is the birds' increasingly synthetic diet. With their stomachs stuffed full of plastic odds and ends, the chicks may not be able to eat or drink, or even recognize that their bodies need food or water. In one study funded by the Environmental Protection Agency, researchers found that the chicks that died of dehydration or starvation had twice as much plastic in their stomachs as those that died of other causes. Then too, plastic is sometimes a direct cause of death, as when a chick swallows a piece of plastic sharp enough to puncture its stomach lining or so large it blocks its esophagus.

Klavitter can't usually tell if a bird is being affected by plastic while it's still alive. The birds don't typically show signs of distress, and since their stomachs are designed to deal with indigestible items, such as squid beaks, they often will just regurgitate plastics they've swallowed. But once he spotted an adult albatross trying to cough up a piece of plastic that was apparently stuck in its throat. He caught the bird and massaged its throat to carefully extract what turned out to be the long white handle of a child's pail. The Laysan albatross is just one of more than 260 species of animals in the world that are being killed or injured by plastic. And, said Klavitter, "what we see in the albatross is happening further down the food chain." Fish large and small and even dime-sized jellyfish are ingesting plastic. "It's pretty spooky how far down it goes."

I've looked at photos of dozens of dead Laysan albatrosses — pictures that capture in the starkest way the threat plastics pose to the natural world. Every carcass seems a mockery of the natural order: a crumbling bird-shaped basket of bleached bones and feathers filled with a mound of gaily colored lighters and straws and bottle caps. The birds are dissolving back into the ground; the plastics promise to endure for centuries.

You can chart a direct relationship among the rising production of plastics, humans' growing reliance on throwaway products such as disposable lighters, and plastics pollution of the environment. As British biologist David Barnes wrote, "One of the most ubiquitous and long-lasting recent changes to the surface of our planet is the accumulation and fragmentation of plastics." And it's happened in the space of a single generation, really just since the 1960s, when the era of disposability took hold.

The very qualities that make many plastics such fantastic materials for the human world — lightness, strength, durability — make them a disaster when they get loose in the natural world. The air, the land, the sea all carry the tracks of our dependence on these most persistent of all materials. I see the evidence in my family's backyard compost bin, where every so often a tiny bar-coded plastic label — the sort inevitably affixed to fruits and vegetables today — comes bubbling out of the fresh new soil: *Nectarine 3576; Avocado 2342*. That many of the labels also read *organic* seems particularly perverse.

Plastic bags, packages, cups, and bottles skitter across landscapes around the world, from crowded urban downtowns to remote rural areas, an eyesore for humans as well as a hazard for wildlife. Over the past year, I've found four plastic lighters during my daily walks in Golden Gate Park. Even plastics that are properly thrown away in landfills can cause problems, leaching endocrine-disrupting chemicals such as phthalates, bisphenol A, and alkyphenols that can contaminate the soil, streams, and ground water.

But most worrisome of all is plastic pollution of the oceans, the terminus of all inland waterways, the bottom of the hill from everywhere that humans live. There, far out of sight, plastics are accumulating at a mind-boggling rate, settling on the ocean floor at all depths and aggregating across vast stretches of open sea, as in the North Pacific area where that castaway lighter was found. No one knows for sure how much plastic is in the world's oceans — I've seen estimates ranging from 13,000 to 3.5 million pieces of plastic per square kilometer. One expert estimated as much as 1.6 billion pounds of plastics ends up in the oceans each year, which is about the same amount of Atlantic cod fished from the seas annually. A key question researchers are now asking is whether that plastic — and the toxins it can carry — is getting into the food chain. Is our plastic trash winding up back on our dinner plates?

Researchers first began noting the presence of plastic in the ocean in the mid-1960s. The problem has increased exponentially along with rising plastic production, which has grown twenty-five-fold in the last half century. During that time, the volume of plastic fibers in the seawater around the British Isles rose two- to threefold, surveys show. Off the coast of Japan, the amount of plastic particles in the ocean rose even more sharply, by tenfold during the 1970s and 1980s and then tenfold again every two to three years during the 1990s. While in some areas, such as the North Pacific, the problem continues to worsen, studies suggest that in other places it may be leveling off. Archives of trawls taken along the East and West coasts of the United States, for instance, have not found increases in recent years.

Still, considering the vast quantities of plastic present in the seas, this may be the most intractable and disconcerting form of plastic pollution. As the BP oil spill in the Gulf of Mexico demonstrated all too tragically, damage deep in the ocean is difficult to repair. How do you even begin to clean up an environment covering 70 percent of the earth's surface; a vast, lawless wilderness that belongs to everyone and, hence, to no one? It's a classic tragedy of the commons, and a dire one, since this commons is the "blue heart of the planet," as oceanographer Sylvia Earle called it, the wellspring of most of the oxygen in the atmosphere and home to more species of plants and animals than any other habitat.

The oceans have absorbed humanity's castoffs for centuries, but "we're reaching a tipping point," warned Richard Thompson, the dean of marine-debris research. "It's going to be difficult to remove the plastic debris that's already in the environment, if not impossible. With the exponential growth we're seeing in plastic production, in ten or twenty years we're going to have a serious problem unless we change our ways."

The disposable lighter is an icon of the throwaway mentality that began to take shape in the years following World War II, when the technology that helped the Allies win the war took aim at the domestic front. Disposability wasn't an entirely new concept: when paper became cheap, in the nineteenth century, throwaway paper shirt collars came into vogue, and stores began handing out paper shopping bags. Still consumers mostly assumed the things they bought could be used over and over again and if broken could be repaired. The new materials coming out of World War II challenged that assumption by their very nature. Plastics weren't something people could make or fix at home. How could you patch a cracked Tupperware bowl? Was it even worth the bother?

In the immediate postwar years, plastics began replacing traditional materials in durable goods. But it was clear that consumers would buy only so many cars and refrigerators and radios. The industry recognized that its future depended on developing new kinds of markets, and the steady innovations in polymer science were paving the way. The market for short-lived applications was "rosily astronomical," as the trade journal *Modern Plastics* crowed. Or as a speaker at a 1956 conference bluntly told an audience of plastics manufacturers: "Your future is in the garbage wagon."

Pretty soon all those durable long-lasting materials developed for the hardships of war were being turned to ephemeral conveniences of peace. The wonderfully buoyant and insulating Styrofoam that the U.S. Coast Guard had used for life rafts found new life in picnic cups and coolers; the vinyl-based compound Saran, which had proved so useful in protecting military cargo, was redeployed to the short-term protection of leftovers; and polyethylene's extraordinary capacity to insulate at high frequencies was sidelined for a new career bagging sandwiches and dry cleaning.

Initially, such products were a tough sell — at least to the generation that had come through the Depression and wartime scrap drives with the mantra "use it up, wear it out, make it do, or do without." The ethos of reuse was so deeply ingrained that in the mid-1950s when vending machines began dispensing coffee in plastic cups, people saved and reused them. They had to learn — and be taught — to throw away.

The lesson was quickly absorbed, driven home by an ever-expanding array of new disposable products from lobster bibs to diapers (which some pundits suggested were responsible for the postwar rise in birth rates). *Life* magazine celebrated what it dubbed "Throwaway Living" with a photo that showed a young couple and child with their arms raised in exultation amid a downpour of disposable items — plates, cutlery, bags, ashtrays, dog dishes, pails, barbecue grills, and more. Cleaning the nondisposable versions would have taken forty hours, *Life* calculated, but now "no housewife need bother." No wonder the young mom looked so happy! We learned to throw away so well that today half of all plastics produced go into single-use applications.

The disposable lighter arrived in this tsunami of disposables washing over the marketplace, an emblem of the changing mindset. Procuring a match or refilling a butane lighter was hardly an onerous task. Still, even such slight burdens could be lightened through disposability.

Disposable lighters began to be popular in the United States in the early 1970s, about the same time the first plastic soda bottles debuted and a few years ahead of the plastic shopping bag. The lighter was the brainchild of Bernard DuPont, a member of a venerable French family that had been making and selling luxury leather and metal goods for nearly a century, since the Franco-Prussian War. His company made high-end lighters that used replaceable fuel cartridges. DuPont was holding one of those cartridges one day in the early 1960s when it suddenly hit him: why not just outfit the cartridge with a simple striking mechanism, encase it in plastic, and market the result as a low-cost lighter that could be tossed once the fuel ran out? DuPont introduced the product in France in 1961 and in the United States a few years later, calling it the Cricket.

Whether they were puffing away on Gauloises or Marlboros, smokers loved "the attractive toss-away," as the *New York Times* called it. The Cricket was, said the *Times*, "as much a symbol of contemporary America as the durable Zippo had been for the sturdy era of World War II." That caught the attention of two other companies already well versed in marketing throwaway goods: Bic, which brought the world the disposable ballpoint pen starting in 1952, and Gillette, maker of the first disposable razorblade. Each saw in the lighter another product that, like the pen and the razor, could be produced inexpensively and sold through the convenience marts, self-serve groceries, and drugstores that had sprouted across the postwar landscape. Gillette bought Cricket and expanded its production, while Bic introduced its own lighter. Throughout the 1970s the two companies duked it out in a famously brutal marketing fight. But Gillette's chirpy bug logo wasn't much of a match for Bic's suggestive "Flick my Bic" campaign, and in the early '80s Gillette threw in the towel, ceding the market to Bic. By then, worldwide annual sales of disposable lighters had risen sixfold, to more than 350 million lighters. Smokers — and even nonsmokers — were hooked on the convenience offered by disposables. In hindsight, it's amazing how quickly consumers were willing to transfer affection from matches, which were often free, to these smart-looking fire makers, which people had to pay for.

A lighter might seem an anachronism at a time when smoking has gone the way of the three-martini lunch. Yet while smoking rates in the United States and Western Europe are dropping, in many areas of the world, particularly Asia, the former Soviet Union, and parts of Africa and Latin America, cigarette smoking is on the rise. "The global tobacco epidemic is worse today that it was fifty years ago," the World Health Organization lamented in a recent report predicting that at current rates, the number of smokers worldwide would rise nearly 60 percent by 2050.

For those in the lighter business, that's good news. Bic now has markets in 160 different countries and sells *five million* disposable lighters a day. And that's just Bic — that's not counting sales of un-branded disposable lighters, many of which are manufactured in China. In exports alone, China sold more than $700 million worth of lighters in 2008.

Such volumes help explain why disposable lighters are still a common form of debris. Indeed, the disposable lighter is an even more insistently throwaway product than many single-use items that can be reused or recycled. A disposable lighter exists for the sole purpose of igniting a few thousand flames. Once the fuel cartridge is depleted, the lighter's useful life is over. It cannot be used for anything else; it cannot be recycled, because of the fuel. It can only be thrown away.

The plastic that encases a Bic disposable lighter is a tough cousin of acrylic that was developed in the 1950s and is sold by DuPont under the trademark Delrin. It's a plastic known for its strength, hardness, friction resistance, and imperviousness to solvents and fuel, qualities that make it, DuPont boasted, "a bridge between metal and ordinary plastics." That metal-like ability to contain fuel is why Bic chose it for its lighters. It's a plastic built to endure the toughest abuse.

So what happens when one of these used-up Delrin cartridges is carelessly tossed onto the ground or swept out to sea? For help with this question I turned to Anthony Andrady, perhaps the world's leading expert on how plastics behave in the environment. He literally wrote the book on the subject: *Plastics and the Environment,* a 762-page tome respected by industry and environmentalists alike. Trained as a polymer chemist, Andrady became interested in what he called "the plastics disposal problem" in 1980 when he was visiting his homeland, Sri Lanka. He went walking along beaches where he had played as a child and was dismayed to find them littered with plastic bags and wrappers and other debris. He recognized there was a problem in the industry's paradoxical goal of making materials that were both durable and disposable.

Natural materials such as wood or paper melt away through biodegradation, a process that requires the involvement of microorganisms that can disassemble the molecules and cycle their parts back into carbon and water. But as Andrady noted, it took millions of years to evolve the microbial chop shops that can dispatch a tree or a puddle of crude oil. Plastics have been around scarcely seventy years — nowhere near long enough for the evolution of microbes capable of dismantling these huge and complex long-chain molecules.* Instead of biodegrading, most plastics photodegrade, meaning they are broken apart by the ultraviolet radiation in sunlight. As Andrady explained, UV rays fray and fracture molecular bonds, breaking the long polymer chains into smaller sections; the plastic loses flexibility and tensile strength and begins to break apart. Plastics manufacturers routinely add antioxidants and UV-resistant chemicals to slow the process down, one reason why the breakdown rate can vary from product to product.

In any event, the process is not quick. On terra firma, the plastic case of my lighter would slowly photodegrade: within about ten years the shiny coat would dull, and the case would get brittle and crack, fragmenting into smaller and smaller pieces until it became a fine powder of Delrin molecules. Eventually, the long molecules would fracture into small enough sections that microbes could biodegrade them. How long would that take? Decades? Centuries? Millennia?

---

* Plastics that are amenable to biodegradation do exist. But outside of some niche applications, like compostable trash bags and film, these aren't in wide circulation.

Andrady can't say. All he knows for sure is it's an incredibly slow process, so slow that he called it "of little practical consequence."

In the ocean, that process slows to a standstill. Andrady has submerged hundreds of samples of different plastic materials in seawater for long periods and has found that none easily decompose. His research suggests that in a marine environment, polymer molecules are virtually immortal. Which means that unless it's been beached or removed, every piece of plastic that has entered the ocean in the past century remains there in some form or another — an everlasting synthetic intrusion in the natural marine ecology.

At first the castaway lighter would bob about on the waves — like about half of all plastics, Delrin floats. (So do the common plastics polyethylene, polystyrene, polypropylene, and nylon. PET, the plastic used in soda bottles, sinks like a stone, as do vinyl and polycarbonate.) UV radiation would take its toll, but less powerfully than on land; the cooler temperatures of seawater slow down photodegradation, and the lighter would quickly be coated with algae and other "fouling organisms" that block UV rays.

Whether the lighter stayed close to shore or drifted far out to sea, it would be buffeted by waves, and this also breaks plastic objects into pieces. Eventually the lighter, or its fragments, would become so weighted with algae, barnacles, or other foulants that it would sink, joining all those other plastic things that were denser than water. In the icy, pitch-black, nearly oxygenless pit of the ocean, there is absolutely no way for nature to break down polymers. Instead, on "the sea floor, particularly in deeper and still waters, they are doomed to a slow and yet permanent entombment," wrote Murray Gregory, a New Zealand geologist who has been tracking the issue for thirty years. One researcher reported diving down more than two thousand meters near Japan and encountering plastic shopping bags drifting "like an assembly of ghosts."

The impact of these submerged plastics is still unknown. Experts fear a seabed covered in plastic (and all the other trash that's come to rest at the bottom of the ocean) could reduce oxygen levels in the ocean depths, choke organisms that live in sediment, and even upset the exchange of oxygen, carbon dioxide, and other gases between water levels that is fundamental to the ocean's chemistry. Yet no one knows just how much plastic is on the ocean floor. The best we can do is extrapolate from what we see on the shore.

Kehoe Beach is a fairly remote place by urban standards: about two hours north of San Francisco, near the end of the long peninsular finger that forms Point Reyes and then a mile-long hike through a cattail marsh and down an old creek bed to the ocean. It's a place of wild natural beauty, but I was heading there for the unnatural stuff that routinely washes up on the beach. Its location, near where the Bay empties out into the open sea, makes Kehoe a magnet for ocean-borne plastic debris, what the Bureau of Land Management calls with bureaucratic understatement "matter out of place."

Most of that out-of-place matter was originally discarded on land. Only about 20 percent comes from ships, and that amount has probably decreased since 1983, when an international treaty banning ocean dumping went into force. At Kehoe, plastic debris starts washing up after heavy winter storms have flushed out to sea all the tossed and lost detritus that's been flitting down streets, blowing across fields, gathering in storm drains, and accumulating in inland waterways across the Bay Area.

I'd been told about the beach by Judith Selby Lang and Richard Lang, a husband-and-wife team of beachcombing artists who have been collecting plastic debris from Kehoe for more than a decade. Their first date was a hike along the beach, where they discovered they shared a love of making art from plastic trash. For their 2004 wedding—at Burning Man, where else?—Selby Lang fashioned her dress from white plastic bags and decorated it with bits of white plastic culled from the beach.

The couple estimates they've pulled more than two tons of stuff from the mile-long stretch. This is actually not that much compared to famed junk beaches like Kamilo Beach, on the southern tip of Hawaii's Big Island. There, converging currents throw up so much debris that cleanup crews have hauled out fifty to sixty tons at

a time—much of it derelict fishing nets and lines. Such gear is a serious threat to marine animals, and the problem has escalated since the
1950s, when fishing fleets began switching from degradable natural
materials to long-lasting nylon.

The couple aren't trying to preserve their beloved beach. "We can't
possibly clean it," said Selby Lang. "We say we're curating it." They're
using their beach finds to create art that sounds the alarm about all
that matter out of place. They scour the beach for, as Selby Lang put
it, "things that show by their numbers and commonness what is happening in oceans around the world." They then assemble them into
sculptures, jewelry, or photo tableaux: a wreath of children's bar-
rettes, or a display of deodorant roller balls—known as Ban beans
in beachcombing circles—or a grid of dozens of lighters in different sizes, shapes, and colors arrayed in orderly rows. The pieces are
arresting. They have an abstract beauty that draws the eye, and an
emotional impact that hits as you recognize objects that once passed
through your hands, such as the red sticks in one flag-like design that
I realized on closer inspection were the spreaders from the cracker-
and-cheese snacks I used to buy for my kids' lunches.

The leaden skies were threatening rain on the day I visited Kehoe
Beach. I zipped my jacket tight, turned my eyes to the ground, and
started walking. It took a few minutes to recalibrate my inner treasure hunter, to make myself ignore the pretty shells and stones and
cables of kelp and focus instead on all the junk. As my viewpoint
adjusted, I realized the beach was covered with plastic castaways that
had clearly come from all over the Bay Area. There were black rubber tubes used by oyster farmers in nearby Tomales Bay; green chains
used to stake grapevines in Napa Valley, some thirty-five miles to the
east; shotgun waddings from inland shooting ranges; nibs of escaped
balloons; hanks of nylon fishing rope; and, of course, the litter clas-
sics, such as bottles, bottle caps, plastic spoons, food packages, and
a few plastic bags. I pulled half of a green monobloc chair from the
sand and soon spotted not one but two plastic lighters, each rusty
around the metal top but still as brightly colored as a circus tent.

Plastic makes up only about 10 percent of all the garbage the world produces, yet unlike most other trash, it is stubbornly persistent. As a result, beach surveys around the world consistently show that 60 to 80 percent of the debris that collects on the shore is plastic. Every year, the Ocean Conservancy sponsors an international beach-cleanup day in which more than a hundred countries now take part. Afterward, the group publishes a detailed inventory of every item of debris that's been collected. The list itself is a powerful testament to the degree to which plastics serve as "the lubricant of globalization," in the words of ocean activist-researcher Charles Moore. But what's also striking is the uniformity of what's collected. Whether they're working a beach in Chile, France, or China, volunteers inevitably come across much the same stuff: plastic bottles, cutlery, plates, and cups; straws and stirrers, fast-food wrappers, and packaging. Smoking-related items are among the most common. Indeed, cigarette butts — each made up of thousands of fibers of the semisynthetic polymer cellulose acetate — top every list. Disposable lighters aren't far behind: in 2008, volunteers collected 55,491 beached lighters, more than double the number collected just five years earlier.

If nothing else, the detritus collected each year is testament to the degree to which the whole world is becoming addicted to the conveniences of throwaway living. But to really appreciate the toll that this is taking on the planet, you have to head away from the coast and out into the deep reaches of the ocean.

In 1997, Charles Moore, a California-based sailor, was returning home from Hawaii after a race and decided to try a new route that would take him through the northeastern corner of a ten-million-square-mile area known as the North Pacific subtropical gyre. The gyre is a huge oval loop that spans the Pacific and comprises four powerful currents that move from the coast of Washington to the coast of Mexico to the coast of Japan and back again.

On that sunny August day, Moore steered his boat into a remote part of the gyre sailors normally avoid. The winds there are weak, the fish are few, and a perpetual mountain of high pressure hangs overhead, pressing down and making the currents spin in a slow, clockwise vortex, like water circling the drain in a bathtub. Except that here, the vortex never runs out. A lifelong sailor, Moore was used to seeing the odd fishing buoy or soda bottle off the side of his boat. But he'd never seen anything like what he encountered in the vortex. "As I gazed from the deck at the surface of what ought to have been a pristine ocean, I was confronted, as far as the eye could see, with the sight of plastic," he later wrote. For a full week, he wrote, "no matter what time of day I looked, plastic debris was floating everywhere: bottles, bottle caps, wrappers, fragments."

This was the feeding ground of the Laysan albatross.

Moore's find wasn't news to those who study the ocean's currents. Curtis Ebbesmeyer, a Seattle oceanographer, has made a career of tracking flotsam, debris, and the contents of cargo containers lost at sea, such as rubber ducks and sneakers, to better understand the movements of the ocean. He found that debris from North America and Asia is caught up in the gyre currents where it can circle the northern Pacific Rim for decades. But some gets spun into the center, where there is neither wind nor the strong arm of a current to push it back out; it gets trapped. The technical term for the area is the North Pacific Subtropical Convergence Zone, but Ebbesmeyer was the one who bestowed the more colorful name — and the one that has stuck — the Pacific garbage patch. (There's also another debris-dense convergence zone at the other end of the gyre, in the western Pacific near Japan.) To Moore, *patch* didn't begin to describe what he was seeing: an area that he then estimated was about the size of Texas and swimming with three million pounds of debris — the amount deposited every year in Los Angeles's largest landfill.

That detour through the gyre changed the direction of Moore's life. He quit his furniture-refinishing business and turned his attention full-time to researching and documenting the plasticization of the oceans. His alarming dispatches from repeated trips back to the gyre helped bring public attention to the problem. But that awareness, unfortunately, has been shaped by a host of misperceptions, some fed by Moore's initial descriptions.

By now the plastic vortex has taken on an almost mythic quality in the public imagination. In news reports and the blogosphere it is often portrayed as a huge floating island of trash or, as the *New York Times* recently called it, an "eighth continent." When Oprah Winfrey did a show about it — which was hailed by marine-debris activists as a sign that their issue was finally getting the recognition it deserved — she showcased photos of a messy swamp of bottles and bags and wrappers.

Yet the images are a far cry from the reality. The vortex isn't filled with floes of debris. Instead, as voyagers there have discovered, it's a place of singular beauty, where on a calm day the waters are a clear, fathomless cerulean blue, and at night the surface shimmers with ghostly green bioluminescent trails traced by fish coming up to feed. It's not uncommon to come across bobbing detergent bottles, runaway buoys, or the occasional car-sized clumps of drift nets packed with all types of smaller debris, from toys to toothbrushes. But they're not omnipresent. Doug Woodring, a Hong Kong businessman and ocean activist, spent a month in the vortex during the summer of 2009 as part of a scientific expedition, and he was struck by the absence of plastic bags. He's used to seeing them all over the Hong Kong harbor, but he saw none in the vortex; it's so far from land they would have long since been sunk or smashed to smithereens by the ocean's currents. Mostly what he saw was something far more insidious: gazillions of tiny bits and pieces suspended, like the flakes in a snow globe, throughout the water column, from the surface to the visible depths. Researchers on his ship trawled the waters twice a day with surface skimming nets, and every single time the nets were brought up, they were covered with this plastic confetti. A floating trash island would be a far easier problem to take care of. Ironically, that horrific image actually understates the problem, making it sound containable, amenable to an open-sea version of a beach cleanup.

But unlike a beach, the vortex "is not a static environment," said Seba Sheavly, a Virginia-based consultant who has worked on marine-debris issues since 1993. "It changes with the seasons. It moves. It's very dynamic. To call it a 'garbage patch' insinuates it has boundaries and can be measured. It cannot." In a place as vast as the Pacific

Ocean, said Sheavly, the concentration of debris in the vortex is on the order of a few grains of sand in an Olympic-sized swimming pool.

Sheavly serves as adviser to Project Kaisei, a group that was originally organized by Woodring and other activists in 2009 with the admirable, if naive, goal of using ships equipped with nets and scoops to "capture the plastic vortex." But the group's leaders quickly recognized that there was no way to capture any but the biggest pieces of debris, such as drifting nets. And one scientist warned Woodring that trying to pull out all the tiny floating bits could cause more harm than good. "You can't just scoop out all the plastic from the ocean without pulling out phytoplankton and zooplankton species," organisms that are the foundation of the marine food web. "If you ruin that basis, you'll have a tumbling effect, like taking out bottom bricks from a pyramid."

Indeed, the challenge of clearing debris from the ocean becomes even more stark when you realize that the North Pacific vortex is not the only place on the globe accumulating plastic. Gyres and high-pressure vortices are natural features of the earth's oceans. There are at least five, all centered around the thirtieth parallels north and south, regions known as the horse latitudes, supposedly because the windless conditions there so slowed down Spanish sailors that they had to throw their horses overboard to conserve water. One gyre can be found in the North Atlantic east of Bermuda, where converging currents trap huge mats of sargassum weed, creating the Sargasso Sea. Researchers have been finding plastic debris there since the 1980s; during a six-week survey in 2010, researchers picked up forty-eight thousand pieces of plastic from the area. Other gyres circle the South Atlantic and the Indian Ocean, east of Africa. The largest is in the South Pacific, the windless doldrums where Ahab's crew in *Moby-Dick* had to resort to rowing.

Until recently, little was known about the accumulation of debris in any of the gyres. But in 2009 and 2010, at least half a dozen groups mounted trips to parts of the North Pacific and the Atlantic gyres to publicize the problem and gather information, including Project

Kaisei, Moore's Algalita Marine Research Foundation, and the Plastiki expedition, in which eco-activist David de Rothschild sailed a boat built of used soda bottles from San Francisco to Sydney. And a new organization, the 5 Gyres, is gearing up to conduct research trips to the less explored southern gyres as well.

These currents have probably always carried and accumulated human-generated flotsam and trash. But before the age of plastics, the trash consisted of materials that marine microorganisms could quickly break down. Now the gyres are swirling with stuff that, at best, breaks up into small morsels that are too tough for nature to chew. As Midway biologist John Klavitter observed, "It will take decades for all that plastic to get out of the system. Even if people stopped putting plastic in the oceans today, we'll have plastic coming on to Midway for many more years."

The Laysan albatross's proximity to the garbage patch has made it the poster child for the hazards of plastic marine debris. But the birds are hardly the only animals affected by the increasing presence of plastics in the deep ocean. Other sea birds, fish, seals, whales, sea turtles, penguins, manatees, sea otters, and crustaceans have reportedly ingested or become entangled in plastic debris, resulting, as one researcher put it, "in impaired movement and feeding, reduced reproductive output, lacerations, ulcers and death." How many are killed a year? No one can say for certain. The oft-cited statistic — that debris kills a hundred thousand marine animals annually — is a misquote from a 1984 paper about northern fur seals, which estimated that at least fifty thousand were dying from entanglement in lost fishing gear. (And there is no documented basis for the much-cited statistic that marine debris kills one million sea birds a year.)

But even though they do not have a precise toll, researchers have reported significant casualties. Plastic debris has been identified as the cause of injury or death in 267 different species, including 86 percent of all species of sea turtles, 44 percent of all sea birds, and 43 percent of all marine mammals. Researchers have found animals ingesting plastic from one end of the globe to the other, ranging from fulmars, sea birds that scavenge the Arctic waters of the North Sea, to southern fur seals, which inhabit islands near Antarctica. Even animals that we don't know about know of us through our trash: the first reported specimen of a new whale species, the Peruvian beaked whale, was found in 1991 with a plastic bag wedged in its throat.

Plastic debris may also be exacerbating the fragile state of species that are endangered for other reasons. The population of playful monk seals in the northern Hawaiian islands, already down to twelve hundred, is dwindling even faster because of the dangerous ghost nets that entangle and drown them. Leatherback turtles, a species that survived the extinction of the dinosaurs, is now threatened, in part from choking on plastic bags that they mistake for jellyfish; necropsies of dead turtles found since 1968 show one-third had ingested plastic bags. Endangered humpback whales that migrate between Antarctica and tropical waters to the north have been repeatedly spotted towing tangles of rope and other debris.

British biologist David Barnes fears that plastic debris may wreak even broader havoc by contributing to the spread of invasive species. Organisms can easily hitch a ride on a drifting net or cigarette lighter—and such objects may be even more effective vehicles for transporting species around the planet than ship hulls or ballast water, he said. There's more debris than ships in the oceans, and it travels everywhere, reaching places never visited by boats, such as the far-flung Antarctic Ocean islands, where the same flora and fauna have existed for eons, untouched by the rest of the world's restless mingling. When researchers stepped ashore on the aptly named Inaccessible Island, a tiny volcanic upthrow deep in the Antarctic Ocean, Barnes said they found fishing buoys, plastic bottles, and disposable lighters, any of which could be harboring unwelcome émigrés. The first new species to arrive in such a place could have a massive impact on the system. "Plastic is not just an aesthetic problem," he said. "It can actually change entire ecosystems."

But there also are organisms that may be benefiting from the ever-growing flotilla of plastic, underscoring the complexities of how the natural world is coping with the synthetic assault. The oceans are filled with microorganisms—diatoms, bacteria, and plankton—continually searching for surfaces to glom on to. For them, plastic debris may be an incredible windfall—a veritable shower of FEMA trailers, according to David Karl, a University of Hawaii researcher. Karl's research crew was the one that found the Hong Kong lighter floating in the gyre. The lighter, like every other piece of plastic debris they hauled up in their nets, was coated with a fine slime of microbes, including bacteria and phytoplankton—organisms that are essential to the health of the ocean. To his surprise, Karl found that the plants attached to such plastic objects are copious producers of oxygen, churning out even more from their polymer platforms than is normally produced in open ocean. The finding suggests that, at some level, the multitude of plastic debris may be "improving the efficiency of the ocean to harvest and scavenge nutrients and produce food and oxygen," said Karl. Though given all the harm plastics cause, he was careful to say that he was not advocating more plastic in the ocean.

For all the dangers posed by floating bags, castaway lighters, and abandoned nets, the most profound and insidious threat may well be the trillions upon trillions of tiny pieces of plastic speckled across the world's beaches and scattered through its oceans. These itsy bits, collectively known as microdebris, were scarcely on experts' radar until recently. (The first conference devoted to microdebris was held in 2008.) But now they're what many researchers are most concerned about.

For one thing, their presence is increasing, say scientists who have been tracking marine debris over the decades. Microdebris has been accumulating on beaches around the world, even in remote non-industrialized spots such as Tonga and Fiji. On my trip to Kehoe Beach, I was surprised to see that the sand was suffused with tiny pink, blue, yellow, and white shards of who knows what, as well as the smooth opaque beads that I recognized as preproduction pellets. How had industrial pellets gotten onto this rural beach? They might have leaked out of a shipping container carrying pellets overseas. But more likely, they came from a plastics-processing plant somewhere in the Bay Area, where they may have spilled from a storage silo or loading dock or railcar delivery area and then been blown or washed into storm drains and finally swept out to sea, only to be thrown back up on the beach.

The rise in microdebris is partly due to rising plastics production, which leads to an increase in pellets that can get into the environment: they're now thought to constitute about 10 percent of all ocean debris. It's also due to the growing use of teensy plastic beads as scrubbers in household and cosmetic cleaning products and for blasting dirt off ships. (I recently noticed that there are tiny protective beads affixed to the tips of new pens.) But the main source of microdebris is likely macrodebris: the larger pieces of junked plastic that have been fragmented by the sun and waves. Increasingly, experts fear that these bits are just as dangerous to marine wildlife as the lethal necklaces of packing straps and nylon netting that can choke seals and sharks and even whales.

The poster child for this kind of threat will be nowhere near as charismatic as the Laysan albatross. Instead it might be a lowly invertebrate, like the lugworm, a long reddish-brown creature that burrows in coastal sediment.

That's one of the animals Richard Thompson, a marine ecologist at the University of Plymouth in England, works with. Thompson trained to study minute marine organisms, like diatoms, algae, and plankton, but in the past decade his research has come to focus on the impact of minute marine debris. In one series of lab experiments, he fed micro-sized plastic particles and fibers to three different bottom-feeding creatures: lugworms, barnacles, and sand fleas, all of which eat various types of beach detritus. They all promptly ingested their synthetic meals. Sometimes the particles blocked their intestines, which was fatal. But if the bits were small enough, they passed through the animals' digestive tracts without consequence. Another researcher did a similar feeding study with mussels: not only did the mussels eat the plastic bits, but forty-eight days later, the bits were still in the mussels' systems.

It's not just bottom dwellers that are ingesting microdebris. In a

2008 return trip to the Pacific gyre, Moore harvested hundreds of lantern fish, small fry that dominate the middle depths of the ocean and rise to the surface at night to feed on plankton and, it now appears, plastic. Moore found that 37 percent of the fish he sampled had plastic in their guts; one had a bellyful of eighty-three plastic bits, a considerable cargo for an animal scarcely two inches long. The fish are a staple in the diet of the tuna, swordfish, and mahi-mahi caught near Hawaii, which in turn are popular in the diet of the next creatures in the food chain: us.

That's particularly disturbing in light of recent findings about what those plastic bits may contain. Japanese researchers have found that pellets and fragments of certain plastics (particularly polyethylene and polypropylene) act as sponges, sopping up toxic chemicals that are widely present in the oceans, including PCB, DDT — two carcinogens long banned in the United States — and endocrine-disrupting substances such as bisphenol A, fire retardants, and phthalates. Geochemist Hideshige Takada has found that pellets collected from the world's beaches can contain concentrations of chemicals 100,000 to 1,000,000 percent higher than the surrounding water or sediment. Ironically, that's no surprise to scientists who study ocean contaminants; they've long used plastic beads for just that purpose, to pull toxins out of seawater samples. Indeed, Takada argues the pellets can be used to monitor the presence of persistent organic pollutants in oceans around the world.

I packed up a hundred and fifty pellets I collected on Kehoe Beach and sent them to Takada for analysis. The report I got back nearly a year later showed that the pellets contained small amounts of pesticides, including DDT, the compound that prompted Rachel Carson's book *Silent Spring* and that has been outlawed for all but limited uses since 1972. The analysis also showed that the pellets had "moderate concentrations" of PCBs (ninety-six nanograms per gram), which, according to Takada, was a much higher level than pellets found in Central America or tropical Asia but lower than those from urban areas such as Boston Harbor, Seal Beach in Los Angeles, and Ocean Beach in San Francisco.

Thompson and other researchers fear that these microplastics represent tiny time bombs that could be getting into the marine food chain and working their way back up the long ladder to us. Though there are still more questions than there are clear-cut answers, the early evidence is worrisome. More than 180 species have been documented to eat plastic debris, and the small number of studies to date suggests that chemicals absorbed into those plastics can also desorb into the animals' systems and bodies. Thompson placed lugworms in sediment containing contaminated plastics and found that after ten days there was a higher concentration of chemicals in the worms' tissues than in the surrounding mud, suggesting that the chemicals had leached off the microbits and into the worms. Thompson's colleague Emma Teuten fed sea birds pellets laced with notoriously persistent PCBs and later found traces of the chemical in the birds' tissues and preen glands.

The PCB finding is especially disturbing because once PCBs are ingested, the chemicals migrate to fatty tissues, where they remain. The consequences of that persistence have been sadly demonstrated in the Arctic ecosystem, where over the course of several decades the chemicals have risen up through the food chain from small fish to big fish to polar bears, seals, and whales, and finally to the Inuit natives, who, owing to a diet rich in the fatty meat of seals and whales and bears, have been found to harbor some of the world's highest levels of PCBs in their blood and in the women's breast milk.

Still, it's complicated to pin down the role polymers play in passing on these toxins that are already so widely present in the environment. For example, University of Connecticut researcher Hans Laufer found alkyphenols — chemicals used in making plastics and rubber — in the blood, tissues, and shells of lobsters. He suspects the compounds may be responsible for a shell-softening disease that has devastated East Coast lobster populations. But how did the alkyphenols get into the lobsters? As bottom feeders, the lobsters might have ingested contaminated plastic fragments or smaller organisms that had eaten contaminated bits. Or they may have absorbed the alkyphenols directly from seawater. The ocean, the seabed, and the coastlines in many parts of the world are already polluted with chemicals. The question, said Thompson, is "How much worse do plastics make it?"

Certainly plastics enable the mentality that makes it all too easy for a person to simply discard an object when it is used up with little thought about the consequences. The era of disposability has fundamentally changed our relationship to the things surrounding us — whether of our own manufacture or nature's. Consider, for instance, the mental and cultural shift involved in adopting something like the disposable lighter.

Disposable lighters were not replacing disposable paper matches only. The tool they were really stepping in for was the refillable pocket lighter — most famously represented by the Zippo, the inexpensive lighter made of chrome and steel that gained an enduring place in America's heart during World War II, when they were standard issue for every GI. From their debut in 1932 to this day, Zippos come with a lifetime guarantee: "It works or we fix it for free." Over the decades, the Bradford, Pennsylvania, company has repaired nearly eight million.

Although Zippos, like Bics, are mass-produced and made from humble materials, there is a thriving collectibles trade in them; not the case for Bics or other disposables. Collectors like the advertising logos and themed images Zippo has always printed on the sides of its lighters. Bic has done the same; it offers limited-edition lighters each year, which are decorated with NASCAR heroes or sports-team logos or nature-themed pictures of wildlife and trees. Still, collectors have little interest. "We don't really consider they are lighters," said Judith Sanders, a member of a collectors club called On the Lighter Side. Ted Ballard, an Oklahoma lighter lover who used his collection of forty thousand lighters to create the National Lighter Museum, sniffed at the idea of collecting Bics. "It came to a pretty sad world when people would accept a plastic lighter as a thing they'd carry in their pocket," he told me. "There's no esteem to it."

Why do people "esteem" their durable lighters but not the throwaways? Technically speaking, there's not a huge difference between

Battle of the Bag

I'S NOT ALWAYS easy to see when a relationship is in trouble. People have been known to cut down their forests, exhaust their local water supplies, and deplete their soil, failing to recognize or understand the natural foundation for human existence. Plastics seemed to promise a new foundation for human life: food comes sliced and diced and packaged in plastic; sports are played on grass made of plastic; homes are wrapped in plastic; and every year brings new time-saving devices and electronic miracles encased in plastic.

Now we've begun to acknowledge there is trouble in this relationship, perhaps deep trouble. But we've been together so long it's difficult to imagine a different world, one in which people determined the fate of plastic, rather than the other way around.

And yet a small but determined group have begun to imagine that world. They've realized that the best way to prevent the oceans from choking on plastic debris is to better manage that debris on land, which means, among other things, curbing human reliance on throwaways. As a start, they've trained their sights on the most ubiquitous of all throwaway items: the plastic shopping bag. The bag may not be any more pernicious than foamed polystyrene cups or picnic a Bic and a Zippo. Both rely on essentially the same mechanism to make fire: fuel released through a valve and sparked by the turning of a flint wheel. But the fact that a Zippo can be refilled and a Bic cannot bespeaks a world of difference. If you can't reuse or repair an item, do you ever really own it? Do you ever develop the sense of pride and proprietorship that comes from maintaining an object in fine working order?

We invest something of ourselves in our material world, which in turn reflects who we are. In the era of disposability that plastic has helped foster, we have increasingly invested ourselves in objects that have no real meaning in our lives. We think of disposable lighters as conveniences — which they indisputably are; ask any smoker or backyard-barbecue chef — and yet we don't think much about the tradeoffs that that convenience entails.

The convenient disposables that *Life* magazine celebrated in 1955 appeared in the picture to be magically suspended midair; readers didn't see the next frame in the photo shoot, the one that would have shown them piled up on the ground. For decades, humans have accepted the illusion of that first photograph — that convenience comes with no cost or consequence. But now we're recognizing that our plastic throwaways do not simply go away. They go *somewhere,* and, in the worst case, they become matter out of place.

forks or carryout clamshells, but it's the single-use item that, more than any other, has aroused popular ire. People the world over are calling for the bag's abolition, from community activists who view it as very nearly the spawn of Satan to the somewhat more staid head of the United Nations Environment Program who contends that "there is simply no justification for manufacturing them anymore, anywhere."

In 2007, San Francisco became the first U.S. city to ban plastic grocery bags, joining dozens of other cities and countries on every single continent that have taken moves to rid themselves of the bags. Inspired by San Francisco, local governments across the United States — from Plymouth, Massachusetts, to sunny Maui — announced their own measures to eliminate plastic bags, as did major retailers such as Ikea, Whole Foods, Walmart, and Target. All told, more than two hundred anti-bag measures have been introduced in the United States, and although the plastics industry has successfully defeated or derailed many of those measures, activists as well as industry insiders predict that eventually the plastic bag as we know it will disappear — at least from grocery stores. (There are countless other types of plastic bags we rely upon.)

It's not hard to see why the bags have become a favorite target. They are virtually without substance — evanescent puffs of polyethylene, transient and yet ubiquitous. They are designed for a brief use but remain with us seemingly forever as a visible and costly source of litter — hanging from trees, plastered against fences, tumbling across beaches — as well as a potential threat to marine life. They do cause real harm, but their symbolic weight is even more significant. They've come to represent the collective sins of the age of plastic — an emblem "of waste and excess and the incremental destruction of nature," as *Time* magazine put it. The bags signify the overpackaged world we all love to hate; they're the totem of all the ways in which "the plastics industry has helped turn us into a disposable society," as one anti-bag activist complained.

Often when we find ourselves caught in a relationship that makes us feel bad or guilty, we want to be rid of it as quickly as possible. Yet in the rush for a quick divorce, we may just find ourselves falling into a rebound romance that is no healthier than the one we just left.

How did we get hooked on plastic shopping bags?

For a century, imaginative entrepreneurs eyed the protean possibilities of plastic and asked: Which natural substances can these wonder materials replace? It was not a question that went uncontested. As the editor of the trade journal *Modern Plastics* observed in 1956, "Not a single solid market for plastics in existence today was eagerly waiting for these materials." Each new plastic product faced "either fearsome competition from vested materials or inertia and misunderstanding in acceptance, all of which had to be overcome before plastics gained a market."

The fight to conquer the checkout counter was part of plastic's long and steady incursion into the general field of packaging. About half of all goods are now contained, cushioned, shrink-wrapped, blister-packed, clamshelled, or otherwise encased in some kind of plastic. Indeed, one of every three pounds of all plastic produced is used for packaging, including the now ubiquitous grocery bag that wraps itself around your fingers at every opportunity. The push into packaging began in the late 1950s, as plastic challenged one after another of paper's strongholds. Soon, sliced bread was being sold in plastic bags, and waxed paper was being replaced by sandwich baggies. Dry cleaners abandoned heavy paper bags in favor of polyethylene sacks.

That last shift, however, sparked a national crisis in 1959, with a flurry of news reports that the new filmy bags could kill: eighty babies and toddlers had been accidentally suffocated, and at least seventeen adults had used them to commit suicide. In the ensuing "bag panic," dozens of communities proposed banning the bags, confronting the industry with its first major threat. The manufacturers of film plastics scrambled to save their fledgling industry, spending close to a million dollars on a national education campaign to warn consumers about the dangers of the diaphanous bags while also developing new industry standards to make them thicker and less clingy. Meanwhile, the head of the Society of the Plastics Industry (SPI) pledged to run newspaper and radio ads "until there is not a mother, father, boy or girl in this country who does not know what a plastic bag is for . . . and what it is not for." The combined measures shut down the calls for bans. As Jerome Heckman, the lawyer who for decades represented the SPI through the bag panic and countless later fights, recalled, "Our job was and should always be to open plastics markets and keep them open."

One of the companies with a huge vested interest in opening new markets was Mobil Oil, then the leading producer of polyethylene film. By the time a young college graduate named Bill Seanor joined Mobil in 1966, the company had already developed an extensive line of substitutes for paper packaging. Its bag-on-a-roll had replaced paper sacks in grocers' produce sections, and its Hefty trash bags had helped alter people's longtime habit of lining their garbage pails with newspaper. Restless for new possibilities for polyethylene film, in the early 1970s Mobil began eyeing one of the most lucrative paper products of all — the retail shopping bag. In fact, said Seanor, the company had already spent years and millions of dollars trying to develop a square-bottomed, stand-up plastic version of the classic brown paper bag. "The conventional wisdom was that you had to have the same thing." But because the copycat bag priced out higher than paper, "it never got off the ground."

Then Mobil officials caught wind of a grocery bag that a Swedish company was distributing in small numbers, mainly in Europe. Its inventor, Sten Thulin, had come up with a design unlike any traditional paper bag. Solving technical problems that had stymied other inventors before him, Thulin devised an ingenious system of folds and welds that made it possible to transform a flimsy tube of polyethylene film into a strong, sturdy bag. In its 1962 patent drawings, the bag looked like a sleeveless scoop-neck T-shirt, hence the name now widely used by the industry: the T-shirt bag.

According to Seanor, who oversaw the company's early foray into the production of T-shirt bags, Mobil executives immediately recognized it was the bag for them. They could see that, unlike Mobil's initial design, this bag had the punch to knock paper from its perch at the checkout stand. Indeed, the bag ultimately proved so popular with retailers precisely because it wasn't like the traditional flat-bottomed paper sack. Thulin drew on the distinctive virtues of polyethylene to create a wholly new kind of bag. Today the bag is so maligned that we forget what an engineering marvel it is: a waterproof, durable, featherweight packet capable of holding more than a thousand times its weight.

Seanor and his colleagues may have been excited about the bag they introduced to the United States in 1976 (the inaugural versions were decorated in red, white, and blue in honor of the U.S. Bicentennial), but shoppers were underwhelmed. They didn't like the way a checkout clerk often licked his fingers to pull a plastic bag free from the rack, or the fact that the bags wouldn't stand up, Seanor remembered. "People would get their groceries, take them out to the car, and they'd fall over, and consumers would be madder than hell." And when shoppers were unhappy, it was grocers who caught the flak.

It was clear to the budding bag industry that to win over consumers, it would have to win over grocery stores first. One trade group, the Flexible Packaging Association, launched a public relations campaign that urged grocers: "Check Out the Sack. It's Coming on Strong." Meanwhile, the bag companies reached out directly to stores with educational programs to help grocers overcome shoppers' distaste for the bags. "We put together training programs that told the store how to actually pack plastic sacks," said Seanor, who eventually left Mobil with a few colleagues to start their own plastic-bag company, Vanguard Plastics.

But the most persuasive factor in the new bags' favor was basic economics: plastic bags cost a penny or two, paper bags cost three to four times as much, and because they were heavier and bulkier, they were more expensive to transport and store. Two of the country's biggest grocery chains, Safeway and Kroger, made the switch to plastic bags in 1982, and most of the other major chains soon followed suit. "Once we started getting the Krogers of the world to change, it was pretty much over," recalled Peter Grande, a veteran of the business and now head of a Los Angeles bag company, Command Packaging. There were still occasional skirmishes with paper-bag makers over regional markets, he said. "But the feeling within the plastics industry was 'this is the future — plastic is going to dominate the landscape.'"

The accuracy of that prediction would come back to bite the industry. Plastic bags were so cheap to produce and distribute that, in the inexorable logic of a free market, they were bound to proliferate. Producers would, of course, make as many as they could sell, and grocers had no incentive to ration them. Purchase a few items at the grocery store, and between double bagging and sloppy packing, you might walk out with a dozen bags. (Which in turn gave rise to a whole new market niche: plastic products to hold used plastic bags. I've got two bag organizers in my broom closet.) By the new millennium, the T-shirt bag had become perhaps the most common consumer item on the planet. Worldwide, people used somewhere between five hundred billion and one trillion bags a year — more than a million a minute. The average American was taking home about three hundred a year. And yet like so much plastic packaging, the vast majority of these bags wound up in the trash — or worse.

When plastic first began to penetrate the packaging market, it was promoted for its durability, not its disposability. The reason babies in the 1950s could suffocate on dry-cleaning bags was that people were holding on to them for other uses — as DuPont had encouraged when it first introduced the bags. What's more, those early bags were pricey. "Keep [your] clear plastic bags . . . Clean inside and out with a few dabs of a sudsy sponge," an outfit called the Cleanliness Bureau advised readers of the *New York Times* in 1956. "Dry the bag promptly and it will stay lovely for many seasons to come."

But it didn't take long for the industry to recognize that disposables were the route to growth, and for a prosperous public to get comfortable with the idea of throwing plastic packaging away. Especially as that packaging multiplied. Today, the average American throws out at least three hundred pounds of packaging a year; Americans' combined mountain of stripped wrappings and emptied containers accounts for a third of the total municipal waste stream.

Plastic grocery bags would become that stream's most potent symbol.

Mark Murray had plastic bags in his cross hairs long before the current wave of anti-bag warriors did.

Murray is executive director of Californians Against Waste, a statewide group that was formed in 1977 to push for the passage of a bottle bill in California. Its mission has since broadened to encompass a range of waste-related issues, from electronics recycling to dairy-farm refuse. One reason California has long been on the leading edge of solid waste legislation is CAW and, by extension, Murray. He's spent his entire career with the group, having joined as an intern in 1987 — a fresh college graduate and political junkie who arrived in Sacramento in search of a job. He had no great interest in recycling, but for someone with an intensely competitive nature, the issue turned out to be ideal. As a reporter profiling Murray once observed: "It provided battles that were winnable — not like saving the whales or shutting down nuclear power." He may have just stumbled across the issue, but recycling — really, the whole megillah of waste reduction — quickly became an obsession. At the same time, he became well practiced at balancing his idealistic goals with the demands of realpolitik. He's a pragmatist who's open to compromise, at times too much so, according to critics to his left.

Now in his forties, Murray has close-cropped hair with a sharply receding hairline and the zero-body-fat frame of a long-distance runner. Indeed, he's a competitive marathoner, and such endurance is a useful quality when one is a lobbyist for a nonprofit with long-range goals. Murray knows what it is to keep pushing ahead with his sights fixed on a distant finish line. He's been hoping to get rid of plastic shopping bags for more than twenty years.

According to Murray, some waste questions are complicated, but not the ones surrounding plastic bags. "The plastic bag is a problem product," he said flatly when we met for lunch one fall day when the legislature was out of session. "I'm not out there suggesting that we should ban every plastic product. But there are some whose environmental costs exceed their utility, and the bag is one of them."

Murray's chief gripe about the bag is not the oft-cited one: that they clog up valuable landfill space. In fact, studies have shown that plastic bags and other plastic trash take up much less space in landfills than paper waste or other materials, in part because plastic can be more tightly compressed. Nor is Murray concerned that plastic bags can "last hundreds of years in a landfill," which was one of the stated reasons for a bag ban proposed in Fairfield, Connecticut. Nearly all trash — no matter the material — can endure in a landfill. Archaeologist William Rathje — self-proclaimed "garbologist" for his studies of landfills — has unearthed newspapers from the 1930s that were as clear as yesterday's edition, and decades-old sandwiches that looked fresh enough to eat.

Murray isn't worried about what happens when bags get into landfills; he's upset that so many don't. Sometimes litter accumulates when a person carelessly tosses a cigarette butt or soda can to the ground, he explained. "But the plastic T-shirt bag often becomes litter *after* it has been properly disposed of. The plastic bags blow out of garbage cans, they blow out of the back of garbage trucks, off transfer stations and off the face of landfills." They are intrinsically aerodynamic. Indeed, they're even more aerodynamic now than they used to be. In response to an earlier generation of environmentalists' concerns about landfill space, bag makers have made them even lighter and thinner.

Yet unlike paper litter, when plastic bags get into the environment, they don't biodegrade. Twenty years ago, Murray staged what he called a press stunt to illustrate that problem: he tacked a bunch of paper and plastic bags onto the roof of a downtown Sacramento building. Sure enough, as the weeks went by, the paper bags gradually melted away, but the plastic bags just slowly shredded into smaller and smaller pieces.

The environmental implications of that persistence is what has driven many activists—especially those concerned with marine debris—to do battle against the bag. Murray, however, is mainly motivated by concerns about waste—the ever-growing impact of our throwaway culture. He's an advocate of zero waste, a concept that has been gaining ground among policymakers over the last decade or so, especially in California, where several state agencies and a number of counties and cities have adopted it. Zero waste is less a concrete goal than a guiding principle for policies designed to dramatically reduce the torrent of trash we now bury in landfills and burn in incinerators. More than a prescription for diverting garbage into recycled goods or compost, zero waste embodies a broad ethic aimed at lightening the load we're imposing on the planet. Zero-waste policies encourage people to reduce consumption while also pushing industry to extend the lifespan of the things we use by designing and producing products that can more readily be reused, repaired, or recycled. At its essence, zero waste is an ethic of resource conservation that would seem thoroughly familiar to our great-grandparents.

The plastic bag is a fundamental affront to that ethic. It's made up of resources that were a hundred million years in the making, and yet its useful lifespan is measured in minutes—just long enough, Murray said, "to get my groceries from the store to my front door." The bag can't be repaired. It's not easily recycled. And the number of times it can be reused is limited. Bags may see double-duty carrying lunch, picking up dog poop, and lining trash cans, but studies show a plastic bag has to be used *at least* four times to mitigate the environmental impact of all that goes into making and disposing of it. Paper bags have similar environmental impacts, but they don't cause long-lasting litter and they are readily recycled. In Murray's zero-waste world, we'd all be carrying reusable bags. But for years he couldn't get anyone, aside from members of environmental groups, interested in attacking the problem of wayward bags.

Then Charles Moore sailed his catamaran into the North Pacific plastic vortex and lifted the curtain on the back end of our throwaway lifestyle. This dystopian vision was especially disturbing to beach-blessed California, a place that sees the ocean as its backyard.

The state's long coastline is a priceless natural treasure—a draw for its forty-six-billion-dollar tourist industry; a rich and diverse fishery; a mecca for surfers, sailors, swimmers, and scuba divers. Eager to safeguard that resource, California's Ocean Protection Council had called for the curbing of all single-use plastic products through fees and bans. In a state of beach lovers, even free-market conservatives like the Republican governor Arnold Schwarzenegger considered the bags a big enough problem they were ready to bid them *hasta la vista.* Bag litter was what turned Schwarzenegger against the bags. "Trash on the beach has always been Arnold's pet peeve. He can't stand it," explained Leslie Tamminen, an environmental activist who worked with Schwarzenegger and was instrumental in assembling a broad coalition that pushed for a statewide ban on the bags in 2010.

How much do the bags contribute to that vast offshore swirl of debris? No one can say for sure. They're a hazard to sea life, but probably not as great a threat as ghost nets or the hard plastic microdebris that's most prevalent in the gyre. Still, bags are colonizing beaches like a new invasive species: volunteers in the 2008 international beach cleanup picked up nearly 1.4 million. No doubt some were cast off by careless picnickers. But surveys suggest that most come to ground miles inland and reach the ocean via storm drains and waterways. A study in Los Angeles County found that 19 percent of storm-drain litter was plastic bags. That's bad news for California coastal communities because most are under a federal order to keep their storm drains clear of trash that can be swept into the ocean. Complying with the mandate has cost Southern California cities more than $1.7 billion since the 1990s. For those communities, bag debris was potentially "a huge financial burden," said Murray. So by the early 2000s, "you had moderate and conservative local government officials in Southern California also clamoring about the problem of plastic bags, not just environmentalists."

For the first time in his long tenure at CAW, Murray saw a real political opening to move against T-shirt bags. If enough local communities took up the bag issue, it would provide ammunition for a fight at the state level. In his experience, that was the way to win the support of the politically powerful grocers and retailers, who much preferred a consistent statewide policy to a patchwork of local regulations. Declaring war on plastic bags suddenly looked politically feasible.

Then San Francisco fired the first shot, setting off a political chain reaction that put the city, and ultimately the entire state, on the frontlines of the bag wars.

San Francisco prides itself on being at the forefront of green policy-making. There are charging stations for electric cars in front of city hall. Residents get tax credits for installing solar panels. City trucks collect used grease from restaurants for the biodiesel-powered fleet of municipal vehicles. The city has one of the most aggressive recycling programs in the country, recycling or composting over 70 percent of its waste and sending less than 30 percent to landfill — the exact reverse of the national ratio. In 2002 city leaders adopted zero waste as a goal, pledging to reduce that 30 percent to nothing by 2020.

City leaders already tilted green, but they had very practical reasons for targeting bags. City grocery stores were handing out 180 million bags a year, and all this clingy, flyaway plastic was wreaking havoc at San Francisco's state-of-the-art recycling plant. People weren't supposed to put the bags in their recycling bins, but invariably people did. At the recycling facility, the bags would come loose, flutter around, and gum up the works. The plant had to shut down twice a day or more so that workers armed with box cutters could manually cut out plastic bags wrapped around the conveyor belts. Plastic bags were costing the facility about $700,000 annually.

There were other costs as well, according to Robert Haley, a long-time staffer in the San Francisco Department of Environment who is in charge of shepherding the city toward its goal of zero waste. In fact, when Haley totted up all the problems related to plastic bags — at the landfill, the recycling center, as litter on the streets and in parks — he estimated the cost to San Francisco at $8.5 million a year, admittedly a tiny part of the city's multibillion-dollar budget, but in Haley's view, a needless expense for the deficit-plagued city. The bags may have been a nice convenience for San Francisco shoppers. But the price of that convenience, Haley reckoned, was higher than the city could afford.

A ban on the miscreant bags was one obvious solution. The city of Mumbai had done it in 2000 after determining that plastic bags clumping in storm drains had dramatically worsened monsoon floods. The city even set up a special police squad dedicated to ferreting out and fining shops and factories that violated the ban. Other Indian cities followed Mumbai's lead with bag bans of their own, as did Bangladesh, Taiwan, Kenya, Rwanda, Mexico City, parts of China, and other places in the developing world.

But rather than an outright ban, Haley—and Murray, who was working with him—were more intrigued by the idea of putting a price on the bags as a way to discourage their use. Their model was Ireland, which in 2002 levied a fifteen-cent fee on plastic bags, a so-called plastax. Within weeks, use of the bags dropped 94 percent and the amount of plastic bag litter decreased significantly. Carrying a plastic bag in Ireland quickly became about as socially acceptable as "wearing a fur coat or not cleaning up after one's dog," noted one reporter. The plastax generated twelve to fourteen million Euros in annual revenue that was dedicated to defraying the program's costs and to supporting a variety of environmental programs. Although the fee wasn't popular at first, the Irish soon accepted it, and one study even found that "it would be politically damaging to remove it."

Such fees serve as a way to make visible the social costs of a product. We may be used to paying nothing for grocery bags, but that doesn't mean they have no cost. Instead the costs have shifted elsewhere—tucked into the price of food, or showing up in taxes that are necessary to deal with the environmental impacts of the bags' production and disposal. Murray and Haley believe fees shine a light on the environmental costs of products, which can ultimately change consumer behavior and promote better product design.

Any type of single-use bag, paper or plastic, has environmental costs. Each drains down finite resources for the sake of an almost trivial convenience. "You pay for everything else in the store," Haley observed. "Why shouldn't you pay for bags?" A fee ought to persuade people to kick the single-use habit altogether and start bringing their own reusable bags.

So with that goal in mind, in 2004 Haley and his staff put together a proposal urging the city to levy a fee on all grocery bags — paper as well as plastic. They set it at seventeen cents, the estimated cost of each bag to the city. City supervisor Ross Mirkarimi, also a zero-waste advocate, was happy to sponsor it.

The proposed fee was controversial — opposed by retirees on fixed incomes, grocers who didn't want to inconvenience their customers, and, of course, the plastics industry, which blasted it as a "tax [that] is going to hurt those who can least afford it." Using the volatile word *tax* to describe a fee that could easily be avoided by bringing one's own bags would be one of the industry's consistent and most effective strategies in skirmishes to come.

While Mirkarimi was trying to muster the support he needed, the grocers and bag makers teamed up for an end run around the city. They went to the state capital to push for statewide legislation to head off San Francisco's fee plan and "keep this issue from going crazy," as one industry lobbyist put it. The result was a law requiring all major grocery stores to offer bag recycling, but another key provision pre-empted cities and counties from imposing fees on plastic bags. One of the best options for dealing with the problems of single-use plastic bags in California had been gunned down.

Murray, to the fury of many environmentalists, helped draft that legislation. Ever the pragmatist, he considered it a transitional measure that would move the state one step closer to getting rid of the bags. Within a year or two, he figured, it would be clear that store-based recycling didn't work and didn't reduce bag consumption, and he could go back to legislators and say, *See, we told you so. Now let's put a fee on the bags or ban them.*

But he didn't anticipate the explosive effect of limiting cities' choices. Local governments don't like being told what to do by state government, especially on a traditionally local matter like waste disposal. "Telling them they couldn't enact a bag fee just motivated the hell out of them to do something," said Murray. "This lit a fire under the San Francisco Board of Supervisors that hadn't existed there before." Mirkarimi called it "the ricochet that became my ammunition."

With that new ammo, he wrote up a proposal to simply ban plastic bags. The formerly reluctant board of supervisors passed it nearly unanimously, and Mayor Gavin Newsom signed it in April of 2007. The new law barred the major grocery stores and pharmacies from giving out T-shirt bags unless they were made of a compostable plastic. (San Francisco is one of the few cities that can handle such plastics, thanks to its extensive composting program.) The stores could still distribute paper bags, leaving San Franciscans free to continue their single-use bag habits. Haley didn't entirely like it, but he rationalized that paper bags could be recycled or composted, and if one ended up on the ground or in the Bay, it would quickly biodegrade. "One good rain and it will go away."

Inspired by San Francisco's example, cities around the country began drafting measures of their own, nearly all aimed solely at plastic bags and most calling for flat-out bans. Lawmakers in a rural Virginia county were motivated by complaints that escaped bags got caught on cotton plants and mucked up the harvesting and ginning equipment. Philadelphia city councilmen were worried that bags were clogging the city's antiquated sewer systems. Residents of a small Alaskan town were moved to action by bags dangling from willow bushes in the tundra. Coastal communities like the Outer Banks counties in North Carolina and the San Diego suburb of Encinitas cited the issue of marine debris.

Strikingly, these political uprisings were entirely local in character. Unlike attacks on other types of plastic products, this battle had no national drive to "ban the bag." The campaigns against PVC and phthalates and bisphenol A have all been shepherded by well-established environmental groups — Greenpeace, Health Care Without Harm, the Environmental Defense Fund, and others — using the media and the Internet to generate public pressure so that even if lawmakers don't respond to consumers' concerns, retailers might. And indeed, while federal regulators are still considering what to do about bisphenol A, big-box stores have stopped carrying baby bottles that contain the chemical. Such coordinated efforts help explain how, to paraphrase *Fortune* magazine, Walmart became the new FDA. But anti-bag initiatives genuinely sprang from the grass roots, proposed by local activists or officials acting more or less on their own — and not always for the most well-thought-out reasons. One industry lobbyist derided it as the *Sixty Minutes* phenomenon: "you see something in the paper and then on TV and it gives you a legislative thought." The popularity of bans was also surely enhanced by what one writer called their "righteous simplicity." Unlike a fee, a ban didn't ask much of anyone — except the plastics industry.

For an industry that's taken plenty of knocks throughout its history, plastics manufacturers were surprisingly slow to respond to the growing anti-bag phenomenon. Bag makers based in California had recognized since the early 2000s that serious trouble was brewing. They could see that the marine-debris issue had very long legs and that sooner or later those legs would start kicking the hell out of their industry. Many actually shared environmentalists' concerns about the plastic vortex and wanted to address what role bags and other plastic products played in creating the problem. They tracked the bag bans and fees being enacted in other countries and watched with alarm the emergence of groups like Southern California's Campaign Against the Plastic Plague, which aimed to eradicate not just plastic grocery bags but all single-use plastic packaging. "This is not a wacko group whose ideas can be shrugged off," the California bag makers warned in their newsletter.

But the American Chemistry Council, the Washington, D.C., voice of the petrochemical industry, seemed oblivious to the growing public concern over marine debris. (Well, not entirely oblivious. In 2004, a spokesman for the group commandeered the Web address, plasticdebris.org, that the California Coastal Commission had hoped to use for a campaign on marine debris.) The ACC could take comfort in industry polls showing that most manufacturers considered the issues roiling California to be irrelevant outside of that state.

"Our industry has been really slow in picking up on the ocean debris problem," said Robert Bateman, a bag manufacturer in Oroville, California, when I visited him a few months after San Francisco enacted its ban. His company makes heavier-gauge bags than the grocery-store giveaways, so he wasn't personally affected by the push to eradicate T-shirt bags. But the growing antiplastics fervor was frustrating to him since he considered polyethylene a far more earth-friendly material than paper. He'd been predicting a backlash against bags for many years, ever since he first started hearing reports about plastic trash washing up on beaches in the early 1990s. He'd worked with Charles Moore to develop environmental standards for containing plastic pellets and had long pressed the big trade groups to start addressing the issue of marine debris, not just for business reasons but for ethical ones as well. "My family was in the asbestos business," he explained. "We learned the hard way that not facing up to issues is not the best way."

The fragmented nature of the industry may partly account for its inability to appreciate Bateman's alarms. The plastics industry is less a unitary world than a collection of planets in their own distinct orbits. The huge, multinational petrochemical companies that make plastic resins operate in a separate realm from the mostly domestic companies that make plastic products. Historically, each group has had its own trade association, conferences, business issues, and political agendas. The resin makers are represented by the American Chemistry Council (ACC), an immensely wealthy trade group with annual revenues of more than $120 million, a staff of 125, four satellite offices, and a roster of issues that reaches far beyond plastics. The products and equipment manufacturers rely on the Society of the Plastics Industry (SPI), a smaller outfit with an operating budget less than a tenth the size of the ACC and a staff of fewer than forty. In the past, the SPI was plastic's prime defender, but in recent years, the group has focused mainly on trade issues and left it to the ACC to serve as the voice of plastics on high-profile issues. Until recently, bad blood between the two groups kept them from cooperating.

Likewise, the world of manufacturers is fractured along provincial lines, with little sense of shared community between, say, a company that injection-molds car parts and one that extrudes grocery bags. So other sectors of the industry didn't feel threatened in the same way that the bag makers did. "We weren't getting any support from the broader industry groups because their ox wasn't getting gored," said Seanor, the Mobil executive who helped introduce the T-shirt bag to America. T-shirt bags may be a hugely visible plastic product, but they make up a minuscule fraction of the plastics business — about $1.2 billion of the $374 billion American plastics market. Why would the rest of the industry rally behind such a flimsy flag?

It wasn't until San Francisco began developing its bag-fee proposal that the major national T-shirt-bag manufacturers — all of whom were based outside of California — finally sat up and took notice. "We said, 'This will not be good for us,'" recalled Seanor. He and Vanguard cofounder Larry Johnson tried appealing for help to the SPI but were rebuffed. Realizing they had to take matters into their own hands, they called a meeting of the country's five biggest T-shirt-bag manufacturers, inviting them all to gather at the American Airlines Admirals Club lounge in the Dallas/Fort Worth International Airport. Attending were executives from Interplast in New Jersey, API in New Orleans, Sunoco in Philadelphia, Superbag in Houston, Vanguard in Dallas, and a gaggle of lawyers. Each company agreed to kick in money to hire a lobbyist and develop a pro-bag campaign. The first year's funding totaled about $500,000, though in later years, various of the companies contributed more.

For the next two years Johnson and other bag executives traveled around the state of California trying their damnedest to quell the rising antiplastics tide. The effort was a full-time job for Johnson until he succumbed to pancreatic cancer in 2007. The group thought it had bought the bag industry some time by securing passage of the state law requiring grocery stores to recycle bags. But that backfired when San Francisco responded with its ban, triggering a wave of like-minded measures. It had taken years for the industry to beat out paper in the packaging market, and now that hard-won dominance was in danger of disappearing.

What's more, bags weren't the only plastic product coming under fire. In 2008 alone, some four hundred pieces of plastics-related legislation were introduced at the local, state, and federal levels, including proposed bans on polystyrene fast-food packaging, phthalate-laden toys, and bisphenol A–containing baby bottles, and even one proposal to classify preproduction pellets as hazardous substances. Plastics had never before come under attack on so many different fronts. "We are at the tipping point," SPI president William Carteaux warned thousands of industry members gathered for the group's big annual meeting in 2009. "Legislation and regulation threaten to fundamentally change our business model . . . We can't continue to fight back just at the reactive stage when things are emotionally charged. We have to take the offensive and react quicker." Industrywide, people were realizing it was time to get serious.

Ironically, by focusing their bans on plastic bags alone, anti-bag activists had unintentionally handed the plastics industry one of its most potent weapons. For the inevitable result of banning plastic bags was that grocers would revert to distributing ones made from paper. Consumption of paper bags shot up more than fourfold to eighty-five million annually in San Francisco following the ban. As some environmentalists already knew—and others would quickly learn—this wasn't doing Mother Nature any favors.

"Paper bags are terrible. Terrible!" exclaimed Stephen Joseph, a Bay Area lawyer representing California bag makers in their fight against cities banning plastic bags.

Joseph is an unlikely crusader for plastics. He's a liberal independent who "hates Republicans," an environmentalist who despises litter, and someone with no prior connection to the plastic industry. He's a contrarian by nature and a litigator by training. But Joseph said his true calling is being "a campaigner," a locution that probably reflects his upbringing in England. "I love fighting for a cause," he explained. He may serve as a gun for hire, but, he insisted, only for those causes that he could sincerely endorse.

Now in his fifties, Joseph is an imposing figure with salt-and-pepper hair, a high forehead, a long nose, and an irrepressible pugnacious streak. He's enjoyed prior success as a campaigner, most famously when he decided to challenge the food industry's use of trans fats. His stepfather died of a heart attack, and Joseph was shocked to discover that his diet might have been what killed him, so he went on the warpath. His stroke of genius was a 2003 lawsuit to block Kraft Foods from selling Oreo cookies to children on the grounds that they were full of artery-clogging trans fats. The suit generated hostile headlines; Jay Leno and Dave Letterman poked fun at it, and the *Wall Street Journal* called him "the cookie monster." Still, he got the last laugh. Two weeks after Joseph filed suit, Kraft announced it was removing the offending fats. He later ran a successful campaign to make all the restaurants in his hometown of Tiburon trans fat free. Other cities in the country have since followed that lead.

Joseph's success caught the eye of some California bag makers. A man who could win public sympathy with a campaign against Oreo cookies was a man who knew how to champion an unpopular cause. But when they tried to hire him, Joseph turned them down. Then he came across a news article in the London *Times* debunking one of the most commonly cited indictments against plastic bags — that they kill a hundred thousand marine animals a year. As the *Times* discovered, that figure was a misrepresentation of a Canadian study that had implicated discarded fishing nets, not bags, in the deaths of Alaskan seals. "I began digging, thinking if that was a lie, what else was?" Stephens said. The more research he did, the more he became convinced that in the case of plastic vs. paper, plastic was getting an unfair rap. Now he's fighting with a convert's zeal. With his typical in-your-face style, he called his campaign the Save the Plastic Bag Coalition. With atypical reticence, he refused to name the coalition's members.

Joseph can cite chapter and verse from the numerous studies that have shown plastic carries a lesser environmental impact than paper. Life-cycle analyses — studies that analyze a product's cradle-to-grave environmental impact — have consistently found that, compared to paper bags, plastic bags take significantly less energy and water to produce, require less energy to transport, and emit half as many greenhouse gases in their production. Author Tom Robbins called the paper bag "the only thing civilized man has produced that does not seem out of place in nature," but that's true only if you ignore the tree-felling, chemical-pulping, intensive-bleaching, water-sucking industrial production that goes into making that natural, potato-skin feel of a brown paper bag. In reality, it's no more natural than its crinkly polyethylene counterpart (though it typically will contain more recycled content). If your top environmental concerns are energy conservation and climate change, plastic is unquestionably a greener choice than paper.

However, life-cycle analyses don't tell the whole story. They do best measuring energy-related impacts, but they have trouble with less easily quantified issues, such as litter and marine debris, the toxicity of materials, and impacts on wildlife.

Perhaps more to the point, data-driven comparisons don't speak to our *feelings* about the two materials — our irrational sense of comfort with the feel of paper bags and our sense of discomfort with plastic's preternatural endurance. The presence of plastic where it doesn't belong — matter out of place — pisses people off. This became very apparent when I accompanied Joseph to a 2008 public hearing on a plastic-bag ban proposed by Manhattan Beach, a small, upscale suburb of Los Angeles that's perched on a hill and overlooks a spectacular stretch of ocean. The town is fairly evenly split between Democrats and Republicans, but everyone there cherishes the beaches — there are more surfers per household than just about anywhere else in California.

Joseph and I arrived early. Even though we weren't dressed for the beach — we were both in suits and carting luggage — Joseph suggested we stroll down to check it out. "Do you see any bags?" he kept asking as we walked along. And he was right — most of the litter we saw consisted of cigarette butts, soda cans, and paper trash. "Gorgeous, isn't it . . . Where are the bags?" Joseph said, staring out over the neatly raked stretch of white sand, either ignoring the fact or unaware that county trucks rake the beaches clean of debris every day.

Like other defenders of plastic bags, Joseph maintained that bag litter wasn't a product problem but a behavioral one: bags don't litter; people do. And as such, he insisted, it made no sense to attack the product for the way it was misused. He pointed to two pizza boxes scrunched up on the sidewalk. "Are we going to ban pizza now?"

Yet Joseph's arguments got no traction in the standing-room-only council meeting that night. No one cared if plastic bags were less pernicious for marine life than abandoned nets or that making them produced fewer greenhouse gases than making paper bags, or that city staff had not thoroughly analyzed the environmental impacts of shifting to paper bags. As one activist testified, "This isn't about global warming. It's about the Santa Monica Bay." Supporters of the measure had two priorities: protecting the area's precious coastline and moving the city away from the use of disposable bags of any kind. Council members said they were starting with plastic bags but hoped eventually to take on paper bags as well. "This is not about the transition from plastic to paper. It's about the transition from plastic and paper to reusable bags," said one. "Changing human behavior takes time."

Every member of the council favored the ban. The minute the last vote was recorded, Joseph turned to me and said, "Lawsuit!"

The fact that paper bags would become the default choice in the city's stores gave Joseph solid grounds for that lawsuit and a winning case in court. He argued — and both a trial court and an appeals court have agreed — that the ban violated a state law that requires cities to prepare a study of the possible adverse environmental consequences of proposed laws. That a pro-environmental regulation could be used to defeat a law aimed at protecting the environment was, one commentator suggested, the legal equivalent of "karma's a bitch."

Throughout California, such lawsuits, or threats of lawsuits, slowed the local drives to outlaw plastic bags, forcing at least a dozen cities — including Oakland, Los Angeles, and San Jose — to back off ban proposals and even withdraw enacted laws. The $50,000 to $250,000 required to prepare a full environmental-impact report is a high barrier for cash-strapped California municipalities. But eventually a group of cities decided to pitch in and pay for a report they all could share. When it was completed, in early 2010, it confirmed what Joseph had been saying all along: paper bags carry many more severe environmental impacts than plastic.

That finding was surprising to some plastic-bag-ban advocates, including Carol Misseldine, director of Green Cities California, the group that commissioned the report. It didn't temper her distaste for plastic bags, but it drove home how the political debate had gone off track: the issue isn't really plastic or paper, she said, but the habit of carrying groceries and other merchandise home in bags designed to be used just one time. "Single-use products have extraordinary environmental impacts in manufacturing, processing, and disposal," she said. "We have to get back to a mindset that relies on durable products."

Roger Bernstein of the American Chemistry Council has understood that all along. He recognizes that the plastic-versus-paper fights are sideshows; the real threat to the industry is the battle against single-use products, the drive to replace disposables with reusables. The push for bans and fees are being driven by "a pure expression of the zero-waste ethic, and total non-choice about reusable bags is the end game," he said disdainfully. "Everything needs to be reused!" Bernstein is vice president for state and legislative affairs for the ACC, which has increasingly taken on the role of chief lobbyist for the plastic bag. I met with him and other ACC representatives at their headquarters in Arlington, Virginia, shortly before the group moved to a new state-of-the-art-green, LEED-certified building closer to Capitol Hill.

Bernstein, in his sixties, is a slight, sharp-featured man with a thick thatch of gray hair and brown eyes magnified behind his glasses. He's been a backroom warrior for the industry for more than thirty years. A former journalist, he started at the Society of the Plastics Industry, then moved to the American Plastics Council, a group formed by major resin makers in the late 1980s, and then joined the ACC when it merged with the Plastics Council in 2000. The ACC had kept its distance from the bag battle, but it took up the fight in early 2008 when the bag makers became overwhelmed by the blizzard of anti-bag measures. The ACC clearly hoped to prevent the furor over bags from snowballing into broader anti-plastics initiatives.

Bernstein divides the politics of plastic into "fear issues" and "guilt issues." Fear issues, he said, are the ones concerned with "environmental self-protection" or with safety questions such as the debate over the potential health risks of bisphenol A. "You have to address those with all the information you can bring to bear on [the chemical's] safety," ideally from third-party sources that will be considered more credible than the industry itself. To that end, the industry has sponsored research studies on suspect chemicals, studies that have a striking tendency to produce results showing the chemicals in question are safe far more often than those conducted by independent researchers do. Bernstein called it delivering information; critics called it sowing doubt.

Plastic bags don't arouse fear, but, as Bernstein recognized, they do play on people's sense of guilt about consumption and the wastefulness of throwaway products. The answer to that is to give people ways to feel all right about single-use plastic products. That means public relations campaigns to remind people about the benefits of plastics and industry-sponsored bills and programs that promote recycling, which he called "a guilt eraser." Recycling assures people that plastic isn't just an infernal hanger-on; it has a useful afterlife. "As soon as they recycle your product," he explained, "they feel better about it." Then they don't want to ban it.

Bernstein learned the methods of guilt assuagement in the late 1980s during an earlier public outcry over plastics packaging. Fears about shrinking landfill space touched off a wave of calls for bans on Styrofoam takeout containers and other visible forms of plastic trash.

In response, seven of the major resin makers, including DuPont, Dow, Exxon, and Mobil, launched a special initiative—a short-term "strike force," as Bernstein described it—to ramp up plastics recycling, at that time virtually nonexistent. The group spent some forty million dollars developing plastics-recycling technology and providing technical help and equipment to communities that wanted to start recycling programs. It was a great boon to recycling, but the commitment was shallow — the support evaporated once the political furor died down.

The heftier and lengthier investment was a $250 million, decade-long campaign of print and TV ads spotlighting how plastics enhanced people's health and safety, with heavy emphasis on products such as bike helmets and tamper-proof packaging. The Plastics Make It Possible campaign succeeded in lifting plastics' favorability ratings, polls showed. People still thought plastics posed serious disposal problems, but they weren't clamoring for bans anymore.

That was also due to the heavy stick the industry wielded, alongside the proplastics carrots. Aggressive industry lobbying succeeded in defeating or gutting hundreds of restrictive bills. "There were no bans, essentially, in all that time," Bernstein recalled proudly. Between recycling, PR, and hardball lobbying, "There were no products that were put out of the marketplace."

The ACC is using that same playbook again. It's launched a major public relations effort, reaching out especially to the millennium generation with a Facebook page, Mylecule (which as of August 2010 had only seven monthly users), a YouTube channel, a Twitter handle, blogs, and sponsorship of art exhibits and fashion shows where the message is "plastic is the new black."

Meanwhile, Bernstein is directing the political combat. He's choosing his battles carefully, focusing on high-profile cities and states to get the most bang for the buck. For instance, the group spent $5.7 million in California during the 2007 to 2008 legislative sessions, when some of the most intense bag debates were taking place, and nearly one million dollars during the months in 2010 when the legislature was considering a proposed statewide ban. By highlighting the environmental problems of paper bags, the ACC succeeded in steering initiatives aimed at banning plastic bags to either voluntary or mandatory store-based recycling programs in New York, Philadelphia, Chicago, Annapolis, and the state of Rhode Island, among other places.

But more recent fights have required the ACC to address the issue of reusability directly — a point where the industry can play to people's mixed feelings about single-use products. In Seattle, for instance, the group waged an aggressive campaign against a 2008 law passed by the city council requiring grocers to charge twenty cents each for either plastic or paper bags — the same approach San Francisco originally wanted to take. Left standing, the law would have marked the biggest victory to date for advocates of reusability. You'd think an ecotopia like Seattle — where the public utilities use goats instead of pesticides to keep down weeds — would be an unlikely place for a plastics showdown. Yet Bernstein and his colleagues realized they had a shot at winning when they got a look at polling conducted by the city. The polls showed that most Seattleites were willing to accept a ban on plastic bags. At the same time, they were unwilling to pay a fee for them at the grocery store. They could live without plastic bags, but not without the convenience of a free one-time-use tote for their groceries. That ambivalence — certainly not limited to Seattleites — offered the ACC an opening.

The group spent more than $180,000 on a successful drive to gather signatures for a ballot initiative to overturn the fee, and then another $1.4 million on the election — the most spent in the city on any election in at least fifteen years. Using the same PR firm that crafted the famous Harry and Louise ads that defeated the Clinton-era health-care reform initiative, the group developed an ad campaign that recast the fee (which citizens could avoid by not buying bags) as a regressive mandatory tax, as in the following radio ad:

Man: You heard there might be a tax on grocery bags, on paper and plastic bags, right?
Woman: Another tax in this economy? . . . But most of us already reuse or recycle these bags.

The campaign maintained that the fee would cost each consumer three hundred dollars a year, which assumed each consumer was buying fifteen hundred bags a year — or twenty-eight bags a week. Whether or not an individual really would purchase so many bags, it was a powerful argument in the midst of the Great Recession, and one that was difficult to counter. There's no easy catch phrase to articulate the logic of making hidden environmental costs visible. What's more, advocates of the fees — groups such as the Sierra Club and the People for Puget Sound — raised just a fraction of the ACC's war chest, leaving them outspent by a margin of fourteen to one. By the time the election took place, no one was surprised when voters rejected the fee.

The ACC followed a similar strategy the next year in California when state lawmakers proposed restricting all single-use grocery bags. The measure was designed to steer Californians toward reusables by banning plastic bags and requiring grocers to charge at least a nickel for paper bags. Given California's political influence, the ACC, and member companies ExxonMobil and Hilex Poly, pulled out all stops to defeat the measure, together spending more than two million dollars on efforts that included peppering the statehouse with donations to key legislators and blitzing Sacramento (where lawmakers lived) with newspaper and radio ads decrying the fee as a regressive tax that would cost Californians more than a billion dollars a year. (The ACC even attacked reusable bags by funding and publicizing research that showed the bags could be a breeding ground for food-borne bacteria.) "Instead of wasting time and telling us how to bag our groceries, lawmakers should be working on our real problems, including a huge budget deficit, home foreclosures, and millions of workers without jobs," the ACC argued on a website that called for voters to "Stop the Bag Police." The arguments may have been beside the point, but even proponents of the ban admired how shrewdly they played to the state's political climate. They made the concern with bags seem silly, as if it were "one of those nanny-government type issues," said Murray. At a time when California was $19 billion in the red and the fractious state legislature was months late in approving a budget, no lawmaker wanted to be seen as a person who banned bags but couldn't manage to organize the state's finances. In the end, the state senate killed the bill, twenty-one to fourteen.

Still, the unprecedented breadth of the coalition that supported the proposed ban — environmental groups, recycling groups, unions, the state's grocers and retailers, and even Schwarzenegger—suggests that the bag's days in California are numbered. Indeed, Murray and other strategists simply shifted their focus, taking the issue, as he said, "back to the locals." In the following months, a number of cities, including San Jose, Los Angeles, and Santa Monica, began moving ahead with plans to restrict T-shirt bags. And unlike the earlier generation of anti-bag measures, these aim to restrict paper bags as well.

California has long been a bellwether state—pioneering the trends that the rest of the country later follows. It's hard to know if that will prove the case with plastic bags, whether the marine debris and waste issues that resonate so strongly in California politics will have the same effect elsewhere in the country. The ACC may have succeeded in squelching most the proposals for bans, but its intense lobbying failed to stop the District of Columbia city council from passing a five-cent charge on plastic bags, the proceeds of which will be used to fund a cleanup of the district's litter-choked Anacostia River. The 2009 measure was promoted with slogans like "Skip the Bag, Save the River." Residents grumbled about it at first, but after nearly a year, according to city officials, consumers and shop owners have come to accept it and are using notably fewer plastic bags. The D.C. experience suggests that people are willing to pay the price of convenience when its true costs are made clear.

Along with the political combat, the ACC continues to push that time-tested guilt eraser—recycling. It has spearheaded a variety of initiatives to spur recycling of plastic bags, from purchasing hundreds of recycling bins to place on California beaches to backing store-based programs. On Earth Day 2009, the ACC announced a more significant commitment: an initiative to manufacture plastic bags with the same proportion of recycled content that paper bags have long contained. Until now, that kind of bag-to-bag recycling has not been widely pursued, since new bags are so cheap to make. The small percentage of bags that are recycled generally go into producing plastic lumber, often used in decking and fences. But the ACC promised that through this new program, bag makers would spend millions to retool their equipment; by 2015, 40 percent of the plastic in T-shirt bags would come from recycled bags. The program would recycle upward of 470 million pounds of plastic, the ACC estimated.

"It is a little too little, a little too late" was Mark Murray's reaction. For even if the initiative fully succeeded, it would recycle only thirty-six billion bags — a mere third of all the bags Americans currently consume every year. Murray and other critics have long maintained that T-shirt bags simply don't lend themselves to the practical and economic requirements of recycling. They're so nearly weightless that it's difficult to gather enough of a critical mass to make recycling them economically worthwhile. Collecting them through curbside programs is tough because the bags are so flight-prone, and the store-based collection programs pushed by the ACC have scarcely raised bag recycling above single-digit rates.

Obviously, it's better for bags to be recycled than thrown away. But the practicalities of bag recycling are largely beside the point. The ACC needs recycling programs to assuage people's guilt about using plastic bags. If we can be persuaded that plastics have a lifetime beyond a one-time shopping trip, then maybe we won't bother to think about the wasted resources the bag literally and figuratively embodies. Thus the new message ACC spokespeople now repeat endlessly whenever called upon to defend plastic bags (or any other single-use plastic product): "Plastic is a valuable resource. Too valuable to waste."

Surely it's no coincidence that this is just the sort of phrase zero-waste advocates use when explaining why they are attacking the plastic bag. To San Francisco's Robert Haley, the bag is the ultimate example of waste, a diversion of valuable nonrenewable resources into an ephemeral product of marginal value. "Plastic should be a high value material," Haley said. "It should be in products that last a long time, and at the end of the life, you recycle it. To take oil or natural gas that took millions of years to produce and then to make a disposable product that lasts minutes or seconds, and then to just discard it — I think that's not a good way of using this resource."

The absurdity of the plastic-bag controversy becomes clear when you consider that people carried things for millennia without the aid of plastic or paper bags. (Be grateful we've passed the era when a bull's scrotum was the bag of choice.) Happily, we don't have to reach that far back to find the bag of the future. A reusable shopping bag can be made of any material — cotton, jute, polyester, nylon, polypropylene mesh, recycled soda bottles, or even thick durable polyethylene. Whatever the material, it will be an improvement over today's giveaways, provided it is frequently reused.

Not all single-use products are so easily replaced. But the fact that the plastic bag can be readily swapped for a sustainable alternative is one reason activists like Murray have put so much effort into the bag wars. That choice at the checkout stand marks an important first step in getting people to think about the environmental consequences of their actions, Murray said. "If you can get people to take this action of bringing their own bags to the store, that's an environmental statement that they're making in their lives," he said. "It's a gateway to environmental activity that I think will spread to other things that they might be willing to do."

As anyone who has tried to quit smoking or follow a diet or commit to a workout routine can attest, it's not easy to change our patterns of behavior, to do things we know, in the abstract, are good for us. So how *do* you encourage people to change their ways, to cultivate habits that are healthier for the environment? Arizona State University psychologist Robert Cialdini has done research for many years on the most effective ways to nudge people toward more environmentally responsible behavior. Surprisingly, the best method isn't to ask people to look inside themselves; rather, ask them to look *outward*, to their peers. "You simply inform them of what the social norm is," said Cialdini. It's not that people don't know littering is wrong or that one should turn off lights when leaving a room. But people forget, grow careless, and need to be reminded, said Cialdini. In one study, he found that the best way to encourage hotel guests to reuse their towels was to leave a card in the room telling them that's what other guests did. That statement had more impact than did cards that told guests they should reuse their towels because it would help the environment, or save energy, or allow the hotel to save money and therefore charge less for its rooms. Another example: Cialdini helped craft a public service announcement designed to encourage Arizona residents to recycle. The PSA said, in essence, Arizonans approve of people who recycle and disapprove of those who don't. It declared recycling was the social norm. Most PSAs move only 1 to 2 percent of listeners to action, according to Cialdini. "Those public service announcements produced a hundred and twenty-five percent increase in recycling tonnage. That's unheard-of."

The frustrating thing about watching the bag wars over the past three years has been seeing how politics reaches for the easy answers, using policies that aren't very effective in making people change the way they think. Fees and public education campaigns help nurture a shared social value of reuse. By contrast, bans may capitalize on and reinforce people's reflexive distaste for plastics without encouraging them to question their reliance on single-use bags of any sort. At least that seems to be the experience in my hometown of San Francisco.

An independent consultant who visited all fifty-four of the city's major grocery stores in 2008 found they all were dispensing paper bags, and in many instances double-bagging whether or not it was needed. True, paper bags may be easily recycled or composted, but San Franciscans are still consuming tens of millions of shopping bags designed for a single trip home from the grocery store. And despite the ban, the city is still awash in plastic bags, since the ban applied only to big groceries and drugstores. Mom-and-pop stores still hand out T-shirt bags, as do produce markets, takeout restaurants, clothing stores, hardware stores, and a host of other retailers. And every morning, rain or shine, my newspapers still arrive in tubular plastic sacks, often double-bagged. I've taken home relatively few plastic bags in over three years, and yet those two bag holders in my broom closet are always overflowing.

For all the shortcomings of bans, they have sustained a public discourse about our single-use habits. And there are hopeful signs that the throwaway mindset is changing. Makers of reusable bags report huge increases in sales; one Phoenix company that makes polypropylene mesh bags saw sales jump 1,000 percent in 2008. ChicoBag, a California company, saw sales of its five-dollar polyester bag triple that year and they have continued to grow. Meanwhile, some makers of plastic bags see a new market opportunity and are retooling to produce heavier-gauge polyethylene bags that are truly reusable.

Recently, I spent one day doing an admittedly unscientific survey at three different grocery stores in San Francisco. While most shoppers were wheeling out carts stacked with paper bags, a small number, maybe two in ten, had their groceries packed in reusable bags — battered canvas bags, heavy plastic totes, or the mesh polypropylene bags that all the major stores in town are now selling for a dollar. Nearly every person in this robust minority said he or she had switched over to reusable bags in the past year or so. I stopped one young woman with five bags in her cart. She said she began bringing her own bags about a year before "just to be environmentally friendly." The reusable bags in her cart looked so new and pristine I asked whether she owned a lot of them. "No," she answered. "I have, like, five, and I always keep them in my trunk. I try not to waste them either."

Closing the Loop

NATHANIEL WYETH OFTEN called himself "the other Wyeth" in deference to his famous artist family: his father, N. C. Wyeth, and siblings Andrew and Henriette. It was clear from an early age that he wouldn't be joining that artistic dynasty. Instead of inks and paints, he was fascinated by gears and gadgets — so much so that when he was ten, his father changed the boy's first name from Newell (his own given name) to Nathaniel, after an uncle who was an engineer. Sure enough, Wyeth went on to train as a mechanical engineer and in 1936 joined DuPont, where he remained for nearly forty years, inventing up a storm in plastics as well as in other materials. It irritated him that chemistry didn't get the same creative kudos as art. A painter need only imagine a picture and then put it to canvas, he pointed out, while a polymer engineer had to conjure entirely new molecules, give them substance, and make them work. As he once told an interviewer, "I'm in the same field as the artists — creativity — but theirs is a glamour one."

The act of imagination that won him a spot in the Plastics Hall of Fame began one day in 1967 with a question: Why does soda come only in glass bottles? Coworkers explained that plastic bottles explode under the pressure of carbonation. Wyeth was skeptical. He bought a plastic bottle of detergent, emptied out the soap, filled it with ginger ale, and stuck it in the refrigerator. When he opened the fridge the next morning, the bottle had swelled so much, he could hardly pull it from between the shelves. Wyeth was certain there was a way to solve what he called "the pop bottle problem." It only took ten thousand tries to find the solution.

Wyeth knew that some polymers gained tensile strength when stretched lengthwise; a polymer used to make a soda bottle needed to get stronger when stretched both lengthwise *and* widthwise. That challenge demanded both a special polymer and a new system for making bottles. Wyeth found both. He used polyethylene terephthalate — a type of polyester — and molded it first into a small test-tube shape that was then inflated into a full-sized bottle; this method caused the molecules to realign and provide a whole new level of structural strength. Here was a plastic bottle that was tough enough to withstand all that pressurized fizz but also safe enough to win approval from the FDA. It was as clear as glass but shatterproof and just a fraction of its weight. Its thin walls kept out oxygen that could spoil food contents while holding in that explosive carbon dioxide. The PET bottle was yet another of those pedestrian plastic products that humbly fulfilled a Herculean set of demands.

Wyeth filed his patent in 1973. Coca-Cola and Pepsi quickly glommed onto the PET bottle, and — as so many plastic stories go — soon there were bazillions. About a third of the 224 billion beverage containers sold in the United States are now made of PET, the polymer also known as #1 plastic, after the resin-code designation introduced by the industry in 1988.

The stunning success of the PET bottle wrought a number of changes that Wyeth, who died in 1990, surely couldn't have anticipated. Soda makers could more easily package their drinks in bottles that were positively Goliath compared to the dainty six-and-a-half-ounce glass bottles that launched Coca-Cola into American iceboxes nearly a century ago. Single-serve bottles swelled to hold three or four times that much; family-sized bottles ballooned to hold up to a hundred ounces. Bigger bottles encouraged bigger consumption. By 2000, the average American was guzzling about fifty gallons of soda pop a year, about double the amount in pre-PET days. And bigger consumption has helped make *us* bigger, say nutrition experts charting worrisome correlations between rising soda consumption and rising rates of obesity and type 2 diabetes.

Wyeth's wonder also enabled the growing habit of consuming on the go. Once upon a time, most beer and soda was consumed in bars and restaurants. Not anymore. Think about how much real estate your average convenience store or corner deli devotes to chilled single-serving bottles of soda, juice, tea, and water. (Mincing no words, the beverage industry calls such stores "immediate consumption channels.") Single-serving drinks are now the biggest part of the beverage industry. The fastest-growing segment of that chug-on-the-go market is bottled water — a controversial product that arguably owes its existence to the PET bottle. Would designer water have become the indispensable twenty-first-century accessory if it had to be lugged around in vessels of heavy, breakable glass?

While wrestling with the pop bottle problem, Wyeth probably didn't give much thought to what would happen to the bottle once the pop was gone. End-of-life issues weren't a big consideration for polymer engineers in his time. Wyeth had come of age in an era when bottlers routinely collected glass empties to wash and refill. By the time he perfected the PET bottle, the beverage industry was already well on its way to abandoning that two-way system. It was a change that had its roots in World War II, when Anheuser-Busch and Coca-Cola shipped beer and soda to soldiers overseas in billions of cans and bottles that the companies knew were never coming back. But the soldiers did return — enamored with the convenience of nonreturnables — and created a demand that helped keep the beer can alive. The shift to nonreturnables also got a boost with the creation of the interstate highway system, which made it possible for the beverage industry to ship goods over much longer distances, eliminating the need for local bottling plants. The introduction of light, unrefillable PET bottles helped seal the changeover to what the industry calls "one-ways."

Plastic bottles intensified the consequences of the shift to one-ways, adding a nondegradable variety of trash to the growing amount of litter carelessly discarded along roadways, beaches, and parks and to the swelling volumes of products and packaging we discarded in our weekly trash. The abandoned soda bottle was an unsightly corollary to the new ethos of convenience, the sort of sour note that might cause you to question a relationship that otherwise seemed like so much fun.

The growing blight of plastic bottles helped shape the sensibility of the nascent recycling movement. More than that, the bottles themselves provided the material basis for that movement to grow. We now have a whole infrastructure of public and private enterprises dedicated to lessening the environmental burden of our expanding waste stream by cycling those one-ways back into use for new raw materials and products. If recycling has an iconic object, it's the PET bottle.

This was partly due to the agility of the PET molecule; it's a polymer that can be easily adapted for a range of after-uses. No sooner did Coke and Pepsi start bottling their sodas in PET than the first PET bottles were recycled into pallet strapping and paintbrush bristles. But Wellman Industries, a longtime textile manufacturer, discovered an even more significant second use for the bottles: the base for polyester fiber. Wellman had been using off-spec industrial scrap to make polyester for years — a strategy that amounted to the company's telling its suppliers: We loved that mistake you made. Can you do it again? The arrival of the PET bottle was a massive windfall, synthetic manna from heaven. Suddenly the company had a source of millions of pounds of inexpensive raw material for clothing, sleeping-bag fill, and furniture. In the 1990s, it teamed up with an old New England wool mill and the outdoor gear manufacturer Patagonia to start turning those used PET bottles into synthetic fleece, launching a whole new category of ecofashion that continues to thrive. Many of the teams at the 2010 World Cup—at least those sponsored by Nike—were wearing uniforms made of recycled PET bottles.

The PET bottle, more than any other plastic product, has succeeded in covering the three bases necessary to a successful recycling loop. It's widely available, thanks to the billions produced each year. It's easy to reprocess. And it has numerous secondary markets. Manufacturers around the world clamor for bottles to make into T-shirts, carpets, and more new bottles. An empty PET bottle is valued by all players in the global recycling network, from street scavengers to multimillion-dollar businesses.

Even so, most are not recycled.

Nationally, we recycle only about a quarter of all PET bottles, like the empty twenty-ounce Diet Coke bottle sitting on my desk as I type, a totem of my daily vice. So of the roughly seventy-two billion bottles produced each year, some fifty-five billion end up being landfilled or littered. That's nearly enough polyester to knit three sweaters for every resident of the United States. It's a collection of energy that could heat and light up 1.2 million households for a year. By any measure, fifty-five billion dumped bottles is an awful lot of waste.

In yet another of plastic's paradoxes, recycling's greatest success story, the PET bottle, is also the emblem of its greatest challenges. We recycle less plastic than any other commodity material—scarcely 7 percent overall, compared to 23 percent of glass, 34 percent of metals, and 55 percent of paper. In short, we're burying the same kinds of energy-dense molecules we spend a fortune to pump from the ground, scrape from mines, and blast mountaintops to reach. How does this make sense? As bag critic Robert Haley pointed out, when we put these precious molecules into products designed for the briefest of uses, we inevitably lose sight of their worth. We forget that an item like a used soda bottle is a resource worth saving, not trash to be thrown away. What will it take to turn that mindset around, to get people to value plastic for more than a one-night stand?

Waste wasn't much of a problem before the twentieth century. Curbside bins and the chasing-arrows symbol may be modern inven-

tions, but people have been recycling and reusing materials throughout history. The preindustrial English were so engaged in recovering
and repurposing clothes, metals, stones, and other materials that one
historian has called the period "a golden age of recycling." Until the
mid-nineteenth century, paper was made entirely of what today we
would call "postconsumer content," namely used rags. During the
American Civil War, rags and fabric were in such short supply that
papermakers imported Egyptian mummies so they could use their
linen wrappings — surely one of the longest recycling loops on record.

For much of the United States' history, Americans produced relatively little trash. Packaging, now one of the largest portions of the
waste stream, scarcely existed. Most food and goods were sold in
bulk. And few people had the resources to be wasteful, as historian
Susan Strasser pointed out in her social history of garbage, *Waste
and Want*. Reuse was a daily habit. Women cooked food scraps into
soup and fed leftovers to the pigs and chickens most households
kept. Old clothes were mended, disassembled for rags, or made into
new outfits. Broken objects were repaired, dismantled for their parts,
or sold to itinerant peddlers, who in turn broke them down and sold
the metals, glass, rags, leather, and other materials back to industry. Children of the poor scoured their environs for useful castoffs
that could be sold, much as they do today in developing countries.
Things that could not be used in any way were burned; for the poor
especially, "trash heated rooms and cooked dinners," Strasser wrote.
The continual cycling of used goods not only kept households going
but also provided crucial sources of raw materials for early industrialization.

These kinds of informal recycling systems began to fade in the
early twentieth century. On the one hand, people were starting to get
more one-way products and packages. And on the other, municipal
waste-collection systems and landfills were introduced by progres-
sive-era reformers pushing to clean up the epidemic-breeding squa-
lor of crowded, turn-of-the-century cities. From then on, the products and materials entering American lives increasingly had just one

final destination: the garbage can. Waste was no longer a source of potential value or opportunity; it was a problem. And it was a problem solved by either digging holes in the ground and burying it or building incinerators and burning it. The worth of waste was measured solely in the tipping fees charged by dumps.

The pendulum began to swing back in the late 1960s, pulled by the emerging environmental movement. Activists were worried about chemicals leaching from unregulated landfills, distressed by growing amounts of litter, and convinced that we were draining the earth's resources at an alarming, unsustainable rate. Invoking that earlier ethos of reuse, thousands of voluntary grassroots recycling programs sprang up in the months surrounding the first Earth Day, in 1970. But the movement didn't really gain steam until the late 1980s, after the New York garbage barge the *Mobro 4000* spent several weeks in 1987 cruising the Eastern Seaboard seeking a place to offload its cargo, and the country became seized by the conviction that it was facing a critical landfill shortage. As it turned out, the *Mobro*'s plight had more to do with its owner's financial troubles than a lack of landfill space. No matter. In the wake of the barge's epic wanderings, modern recycling took off. By the mid-1990s, most states had adopted comprehensive recycling laws and announced recycling targets and goals for reducing the waste headed for landfill. Communities began to incorporate curbside recycling and drop-off centers in their municipal solid-waste programs. Waste haulers built hundreds of materials-recovery facilities, known as MRFs (pronounced *murfs*), to sort the recyclable goods that were collected. New businesses sprouted up that were devoted to reclaiming used materials.

Meanwhile, in 1988 the Society for the Plastics Industry introduced a coding system to help manufacturers and recyclers identify the type of plastic packaging they were dealing with. Hence the tiny numbers you find today on the bottoms of bottles, jars, and other packages. The code was never intended as a promise that the item would be recycled, yet consumers have generally read it that way because the numbers are typically surrounded by a triangle of chasing arrows — the universal symbol of recycling. Given how little plastic is actually recycled, the misconception drives recycling experts nuts.

The resin code covers the six main plastics used in packaging: #1 refers to PET; #2 indicates high-density polyethylene (HDPE), the plastic used in milk and juice containers and T-shirt bags; #3 is polyvinyl chloride (PVC), which in packaging gets deployed in juice bottles, blister packs for electronics, and some cling wraps; #4 is low-density polyethylene, used for frozen-food bags, squeezable bottles, and sometimes cling films and the flexible lids of containers; #5 is polypropylene, the plastic of yogurt containers, margarine tubs, bottle caps, and microwavable ware. The #6 refers to polystyrene, either in a foamed version that's used in egg cartons and takeout containers or in a hard, clear incarnation, which is increasingly being used in clamshell containers for produce, consumer goods, and takeout foods. The last category, #7, was intended as a catchall for any other plastic. In recent years, its association with bisphenol A–containing polycarbonate, found in those hard water bottles, has made consumers leery, tainting the designation for the makers of other plastics. No manufacturer wants a #7 on its product anymore.

The code is such a poor guide to the multiplicity of polymers used in packaging today — which tend to get pigeonholed as #7s — that efforts are under way to revise and expand the number of categories. But back in the 1980s and '90s, the code provided a valuable lingua franca for the recycling infrastructure that was rapidly taking shape. That system itself is now as unreliable as the recycling code, resting as it does on the shaky commitment of municipal governments and people's fuzzily green hopes that by depositing discards in recycling bins rather than the trash, they will reduce their waste.

Most Americans now have access to recycling (though not necessarily through a curbside program). It's the most popular environmental activity we engage in. But does it do any good? Even though I diligently dropped my empty Coke bottles in my blue bin every week, I really had no idea what happened to them after that. As I'd learn, my bottle embarked on an epic journey, and following that bottle would take me to parts of my hometown I'd never before visited, across the globe, and into an economy that is at once both ancient and post-modern.

It was Tuesday morning at 8:20 and I could hear the hydraulic hiss and grind of the recycling truck at the other end of the block. I hurried outside to meet the driver, Bill Bongi, who had agreed to let me follow him on his route.

San Francisco, where I live, probably boasts the strongest municipal recycling program in the country. In its drive to divert as much waste as possible from the city's landfill, San Francisco has made recycling mandatory. Residents are also required to compost food and yard waste. City officials admit they don't really have the resources to enforce this law. Still, because of that larger goal, residents are encouraged to put pretty much any kind of plastics in their recycling bins — the full gamut of #1s through #7s. That means not just the beverage bottles and milk jugs that most recycling programs collect but also yogurt tubs, old toys, flowerpots, toothbrushes, CD cases, disposable cups, and other non-packaging-related plastics less commonly accepted by community programs. There are only a few types of plastics I can't put in my bin: plastic bags, wraps, and film, because they get tangled in the equipment at the recycling plant; Styrofoam, because there are few secondary markets for it; and biodegradable and compostable plastics, because they're designed for composting rather than recycling.

Bongi, who's in his fifties, has worked for Recology, the company that hauls San Francisco's waste, since he graduated high school. Like several other employees of the company, he comes from a family of garbage men. He followed both his father and his grandfather into the business. When I asked if his kids would follow him, he said emphatically, "No, no, no. They're in college." I walked along the sidewalk as he drove slowly up the block, stopping between every two houses to hop out and grab the bins. (He misses the days when the trucks were manned by crews; now the only chance for conversation is if a resident comes out to greet him or ask a question.) There were the blue bins for recycling; green ones for yard trimmings, food scraps, and paper food packaging like pizza boxes that are trucked to the city's industrial-scale composting facility; and black bins, for whatever doesn't belong in the other two. This three-bin system is a key part of the city's ambitious zero-waste policy. Making residents separate their discards into the bins forces them to consider what parts of waste may in fact be fit for useful afterlives. The more city residents put in the green and blue bins, the lower their garbage bills, a "pay-as-you-throw" approach that experts say helps boost recycling rates.

Yet, as in any curbside program, there are intrinsic limitations to how much refuse can be rescued. There tends to be less compliance in apartment buildings, where most San Franciscans live and where responsibility for discards is more diffuse than in a single-family home. It's tough for landlords to check, much less ensure, that all their tenants are putting their recyclables in the blue bin. And curbside programs are, by definition, aimed at collecting things consumed in one place. Which means they don't catch the ever-increasing numbers of beverage bottles that are consumed on the go. Take a look in the trash cans at gas stations, one plastics recycler told me: they're always filled with empty bottles.

I'd made sure to put my week's worth of Diet Coke bottles in my blue bin that morning. But there were surprisingly few plastic bottles in the other bins Bongi emptied. Instead, the contents were mainly newspaper, cardboard, and glass jars. That's because of California's bottle bill: beverage bottles and cans in California can be redeemed for money, and so they are routinely swiped from the curbside bins, Bongi said. "The homeless steal all the bottles and cans." They'll still be recycled, but through the state's redemption program, not the city's recycling system.

I shouldn't have been surprised by the absence of bottles and cans. Months earlier, I had spent a morning with a homeless man named Sean, one of the bit players in San Francisco's unofficial recycling economy. Every day he walks a forty-block route through my neighborhood checking the blue bins for bottles and cans that he can sell to one of the city's eighteen neighborhood redemption centers. Those centers pay for beverage containers by the pound at prices set by the state, which fluctuate depending on the global scrap markets. On the day I spent with Sean, a pound of PET bottles was worth ninety-six cents, a good bit less than aluminum, slightly under glass, but still higher than any other kind of plastic.

Sean was in his late sixties. He was a friendly guy, with a white beard, clear blue eyes, few teeth, a weakness for beer, and a karmic take on his life of homelessness. He was well educated and seemed lucid except when he described the extraterrestrials who were remaking his body to make him stronger and better equipped for life on the street. Still, I believed him when he said that it had been his choice to live outdoors for the past twenty-three years. To his mind, it was all preparation for what would happen when he too was recycled.

"Do you ever want a home?" I asked.

"No," he said, "because the Buddhist teaching is the more you surrender, when you come back [in the next life] you come back bigtime." He had found that whatever he needed tended to come his way, whether that was a shopping cart, a half-eaten sandwich, or a new pair of pants. "I find them on the street. See, there's a whole pile of shoes," he said, pointing to a bag by someone's trash bin. "I just keep my clothes for a day. When I'm finished with them I put them in places where people can recycle them if they want."

Years ago, Sean said, he had no trouble finding enough containers to make fifteen or twenty dollars in a day. But now, like other marginal folks who try to live off recycling bins, Sean found the bins already picked clean by clandestine fleets of cars and trucks that prowled the neighborhoods in the wee hours before the official collectors like Bongi arrived. At the redemption center, Sean watched enviously as an older Chinese couple emptied out dozens of garbage bags' full of bottles and cans that they had collected. Their haul probably didn't come from residential blue bins, the center's director told me, but from local Chinese restaurants and bars with which they had worked out some arrangement. The couple would make more than sixty dollars that day. Sean's take was a mere $5.41 — scarcely enough for bus fare and a beer.

The financial incentives created by bottle bill laws cut two ways. On the one hand, they help collection, ensuring that more bottles get into the recycling stream. California recovers nearly three-fourths of PET bottles sold in the state—almost six times the average of non-bottle-bill states. Yet on the other hand, cherry picking from the curbside bins deprives cities of badly needed revenue for their waste-collection and recycling programs. San Francisco calculates it is losing five million dollars a year to professional poachers, some of whom operate as many as ten trucks. (This money pours into state coffers instead of city ones.) Sometimes Bongi encounters people going through the bins as he is driving his route. Though what they're doing is illegal, he doesn't feel there's much he can do to stop them.

Once he finishes his route, Bongi heads across town to Recycle Central, the city's MRF, the first stop for my empty bottles. MRFs are where the jumbled recyclables collected by a town or county are sorted and baled for sale. San Francisco's MRF opened in 2004 and is a mammoth two-hundred-thousand-square-foot, thirty-eight-million-dollar structure perched on the waterfront. (As is common with waste-related facilities, it's located in one of the city's poorest, predominantly African American neighborhoods.) On the day I went with him, Bongi pulled into the enormous warehouse alongside several other trucks, pressed a button to tilt the truck's back, and then sat quietly staring off into the distance as the truck emptied its load onto the tipping floor. He and the other drivers leave a mountain of recyclables on the tipping floor every day—seven hundred tons is the daily average. Once the trucks drive off, bulldozers rumble in and begin cutting into the sides of the mountain, plowing drifts of discards toward several conveyor belts that rise up into a ziggurat of belts and chutes and massive vacuums where the sorting of materials is done.

Because of San Francisco's reputation for recycling, I had expected to see some fantastically high-tech system. And, in fact, there is some gee-whiz automation. Magnets pull off the steel soup cans, and a device called an eddy-current separator repels aluminum cans and copper, making the cans jump off the lines into different bins. Steep slopes of whirring discs ferry paper and other lightweight goods upward while allowing heavier things to fall onto other moving lines below.

But when it comes to plastics, the system is surprisingly low-tech. Plastics are a challenge for MRFs. There are lots of different polymers and each has distinctive chemical and physical properties, different melting temperatures, and separate secondary markets. "People say plastic is plastic, but a milk bottle is as different from a soda bottle as an aluminum can is from a piece of paper," said Steve Alexander, executive director of the Association of Post-Consumer Plastic Recyclers. Most plastics can't be recycled together, but many look so similar that they are difficult to sort. PET, for instance, is easily confused with PVC—they're both clear and used in the same types of packaging. But just a few PVC bottles in a half-ton bale of PET bottles, or vice versa, can contaminate the whole batch, rendering it unreusable. Even some products that are made of the same base polymer should not be recycled together; a PET bottle that's been blown into shape has a different melting temperature than a PET cookie tray that's been molded through extrusion. Try to combine them and you'll end up with unusable goop. There are machines — optical scanners, special spectrometers, and lasers — that can separate different polymers, but they are hugely expensive, and so few but the biggest regional MRFs have them. That means the sorting of plastics has to be done by hand — which creates jobs, but drives up costs.

And when you've got to get through seven hundred tons of stuff a day, it's not possible to do a rigorous sort. The conveyor belts are moving too quickly for the men and women who work the lines to closely examine each item, much less check its resin code. So they just look for bottles. They pull off bottles made of PET, like the Coke bottles I put in my bin that morning, and of high-density polyethylene, like the containers used for milk, juice, detergent, and motor oil. They also sort the HDPE bottles by color, since a colorless container has more next-life opportunities, and therefore higher value, than one that's sunshine yellow, grass green, or black. (The color in plastics can't be removed in reprocessing, and any dark pigments will render the recycled plastic some shade of gray, at best.)

What about all the other plastic products San Franciscans are encouraged to put in their recycling bins? The yogurt tubs, berry baskets, greasy takeout clamshells, rotisserie chicken trays, half-empty hummus containers, and smeary peanut butter jars that I spotted speeding along the conveyor belts at Recycle Central? The sorters don't even try to separate out this motley collection. The problem isn't the gunky food residue — reprocessors can get even the greasiest jar clean. The trouble is that the value of these plastics in scrap markets isn't high enough to make it worth the time of the seventeen-dollar-an-hour sorters. There's a certain chicken-or-egg quality to the problem posed by these plastics. Most city recycling programs don't collect them because there are no robust end markets, but the end markets can't develop without being assured of a steady supply. Recology's answer to the conundrum is to pack the #3s through #7s together in bales and sell them as mixed plastics, leaving it to others downstream to sort them and find them useful afterlives.

I walked around to the back of the warehouse and watched as huge compressed blocks of the various materials emerge from the back end of the sorting ziggurat. Several bales of smooshed PET bottles were stacked against one wall. Nearby, there were a few smaller colorful bales of mixed plastics and a shipping container half filled with bales of colored #2 bottles.

Up to this point, the scene I'd witnessed was pretty typical of what happens in any MRF. But what happens next can vary widely, depending on where you live and what plastic is involved. If I lived in the Midwest or on the East Coast, for instance, my local MRF would likely sell the bales containing my used Coke bottles to a domestic PET reclaimer, who would shred them into flakes, wash them, and then run those through float/sink tanks to separate the scraps of bottles from the scraps of caps and labels (PET sinks in water, while the plastics used in labels and caps float). The plastic from the caps — usually polypropylene or high-density polyethylene — would be skimmed off and sold to manufacturers to be made into new caps or other products. The PET flakes would go through more cleaning and processing and eventually be sold, as either flakes or pellets, to

manufacturers to make into new PET products, such as polyester fiber, the strapping used to hold packages on pallets, other packaging,
or even new soda bottles. One likely endpoint for my bottles would
be Mohawk Industries, the Georgia textile company that uses tens of
millions of pounds of recycled PET to make carpet.

But on the West Coast there's just one destination for my used
Coke bottles — in fact, for most of the plastics I put in my recycling
bin: China.

China?

Yes, China. For years, it's been cheaper for San Francisco and other
West Coast cities to send their used plastics on a two-week boat ride
to China than to truck them around the country to domestic reclaimers. Indeed, even port cities on the East Coast and in Europe and Latin
America send plastic scrap to China. A freighter can carry far more
tonnage than a truck, and the fuel is cheaper. Also, since America
imports more from China than it exports, ships arrive here packed to
the gills with cargo but return with mostly empty holds. Asian shipping lines historically have been happy to offer reduced rates so they
can make at least some money off that return trip. Surprisingly, by
some analyses, shipping our plastic waste to China is an environmental plus. Using plastic scrap reduces the amount of virgin resin China
has to produce itself, which means fewer greenhouse gas emissions
from its coal-fired factories.

China currently takes about 70 percent of the world's used plastics.
(India, another emerging plastics power, also takes a good chunk.)
Much of that is composed of used PET bottles, almost all of which
are processed by the Chinese into polyester fiber. It takes a lot of fabric to clothe a billion people. Chinese recyclers are able to outbid
their Western counterparts because they enjoy the same advantage as
Chinese manufacturers of plastic products: cheap labor costs.

So I learned when I visited a recycling plant in Dongguan, one of
the factory boomtowns in Guangdong Province. The plant's genial
owner, Toland Lam, also heads the recycling committee of China's
major plastics trade association. Other recyclers I'd contacted were

skittish about being interviewed. But Lam was happy to talk with me, eager to demonstrate that not all of the West's exported trash ends up being sorted and disassembled by impoverished women and children in toxic, illegal operations, like the one highlighted in a 2008 *Sixty Minutes* episode. The program tracked a shipment of computer monitors from a recycling company in Denver to the rural village of Guiyu, where residents' daily exposure to the hazardous metals contained in electronic waste had left most of the children with lead poisoning. (At a recycling conference, two managers of corporate recycling programs confessed to me that such exposés kept them vigilant in tracking their recyclable waste. "I don't want to end up on *Sixty Minutes*," said one.)

Lam was one of the early entrepreneurs in China's recycling industry. In the 1980s, he worked as a petroleum engineer for Arco Oil and Gas and was stationed in Los Angeles when friends in Hong Kong contacted him and asked for his help in connecting them with American companies who might want to unload their plastic waste. Lam saw an opportunity to get back to China and into a potentially lucrative new line of work. At that time, he said, "Nobody knew how to do this. [The scrap] was free. In the early stages, many people made money. Now the market is getting more competitive."

He now owns several plants, including this sprawling facility in Dongguan that recycles postindustrial plastic waste from U.S. businesses — off-spec raw material, shrink-wrap and packaging, and factory-floor tailings. The warehouses — huge open-walled sheds, really — are monuments to excess. I walked through canyons of scrap, feeling like one of the humans in that '60s television show *Land of the Giants*. I passed a towering mound of DVD covers, probably sent from a big-box store; a bale at least twelve feet high of sheets of Cheetos wrappers that for some reason had never been cut into bags. There were ice floes of compressed shrink-wrap; piles of trimmings from disposable diapers. In every case, said Lam, it was cheaper for the American manufacturers to bale up the discards and ship them to him than to reprocess the material themselves.

The reason became clear when we entered one of the sorting sheds.

Several women in aprons were pulling piles of plastic sheeting from enormous bags that were as tall as they were. Some were carefully snipping paper labels off the sheets, a necessary step before the plastic film could be washed and processed into new pellets or flakes. "In China, this labor is cheap. They have many hands," said Lam. "You're talking a couple of hundred U.S. dollars for hiring a person. Can you hire a person for two hundred dollars in the United States?"

"You mean two hundred a month?" I asked.

"Yes, that's what we pay. In the United States, it's two hundred a day . . . It's not worth that kind of labor to separate [materials]. But in China we can do that."

All the workers were migrants who lived in dormitories and ate at the canteen on the leafy, landscaped grounds. The plant was clean and well ventilated; still, the sheds were filled with the smell of hot plastic and the roar of the state-of-the-art machines Lam uses for washing and grinding the scrap back into usable pellets or flakes. Most of the workers were wearing hardhats, and a few had on masks, but I didn't see anyone wearing earplugs or utilizing any other kind of safety equipment. When I asked Lam how he knew what types of plastic resins he was dealing with — a challenge Western reprocessors handle with the use of electronic eyes — he said he could usually tell based on his years of experience. But if need be, he could resort to the lighter test: burning a sample to determine the polymer type from the color and intensity of the flame.

In another building, I watched a man standing atop a large machine and shoveling clumps of plastic into it. These clumps were quickly melted and then extruded out the other end in long gray spaghetti strands, run through a trough of water to cool, and then chopped into small pellets that Lam would later ship back to American manufacturers.

From a global perspective, you could say that businesspeople like Lam are helping to turn the world's one-way waste stream back into a productive loop. "The business world has worked out this solution," said Edward Kosior, an Australian expert on recycling technology.

The neighborhood scrap peddler who took care of waste a century ago is now operating on a much bigger stage. While San Francisco's commitment to zero waste is why I'm encouraged to put broken ball-point pens, used picnic cups, and cracked pails in my recycling bin, China's hunger for those used goods is largely what has assured them a useful afterlife. If I were living in Chicago, a salesman for one of the local waste haulers told me, it would be an entirely different story. Then those items would go straight to landfill. Trying to ship bales of mixed plastics from Chicago to China isn't economical, he explained. And at this point, "in the Midwest, there are no markets for them."

Why have markets failed to develop? China's voracious appetite for America's used plastic is at least part of the answer. All those Asia-bound containers suggest the United States is missing out on an important phase in the plastics economy: reclamation of used resin. While the amount of used PET bottles collected for recycling has steadily increased over the past decade, so has the amount purchased by China. That's taking a bigger and bigger bite out of the used PET that's available for American reclaimers. In 2009, American recycling programs collected a near-record 1.44 billion pounds of PET bottles. But the majority was sold overseas, mainly to China. American reclaimers bought only 44 percent, or 642 million pounds, which was not much more than they bought a decade earlier. The chronic shortage of this material has "stifled innovation and investment" in American recycling, said Alexander, of the plastic recyclers' trade group. It's deterred the development and deployment of technology that could make sorting and recycling more efficient and more profitable.

But while the China factor is undercutting American recycling programs, it's not the only reason why these programs remain underutilized. Recycling rates have been dropping since the early 1990s — even for PET bottles. Experts say the decline is related to the shortcomings of municipal recycling programs. Curbside and drop-off collection programs depend on people's willingness to put their used bottles in the bin. As one analyst put it, "We get people to do it based on a feeling, and that feeling gets you to [a recovery level of] about twenty-five percent." Many people, apparently, feel that recycling just isn't worth the hassle.

To try to entice more people to participate, many municipalities have adopted single-stream collection, where everything can go in one bin. People are more willing to take their bins to the curb when they don't have to separate their paper, glass, plastic, and cans. But in the scrambling of materials, broken glass gets into the paper and plastic; the greasy food residue sullies the paper; and it presents a more difficult, and expensive, stream for the MRF to sort. The law of garbage in, garbage out takes hold: studies show that the more jumbled the materials going into the MRF, the lower the quantity and quality of sorted materials coming out. Bales of PET collected from single-stream programs are often contaminated with paper, broken glass, and crushed cans, making them worth less in resale markets. (One recycling expert told me she'd taken apart a bale of mixed plastics that had been shipped to China and found a vacuum cleaner inside.) This tradeoff between quantity and quality has compromised the success of municipal recycling.

It's also slowed the development of closed-loop recycling systems, in which used plastic bottles can be recycled back into new plastic bottles. Closed-loop systems represent recycling at its most environmentally beneficial. Turning an old bottle into a new one offsets the need for virgin resin, which is ultimately the best way to reduce the resources required to make the things we want. But the FDA has stringent regulations about the quality of recycled plastics that can be used in food-grade products, and the bottles from curbside programs are rarely up to snuff, which is why so many are instead repurposed to make polyester fiber or pallet strapping. Critics call this process downcycling because it's reprocessing the bottles into products that didn't depend on virgin plastics in the first place. Downcycling is the common fate of other types of plastics that get into the recycling stream, such as the grocery bags that are used to make plastic lumber and the milk jugs that are fashioned into landscaping board. Such products may have their own value — there are lots of advantages to plastic lumber—but their production doesn't in any way discourage the ever-mounting production of virgin resin, which is ultimately why we're drowning in plastic waste.

Mary Wood is sure she knows a way to improve the collection and reuse of plastic bottles. She's a huge supporter of bottle bills, such as the ones that now exist in California and nine other states. I ran into Wood at the trade show of a plastics recycling conference in Austin, Texas, in early 2010. Actually, it was hard to miss her—or at least her booth. With a broad banner reading *Plastic Pollution Texas,* it stood out in the hall of plastics-happy business exhibitors. Certainly it was the only one emphasizing the consequences of recycling's failures: one wall featured photos of streams stuffed with empty plastic bottles and a picture of the skull of a pelican with a plastic bottle trapped in its beak. Wood and her booth mate Patsy Gillham were spearheading a drive to pass a law that would put a refundable ten-cent deposit on all beverage bottles and cans sold in Texas, a state with one of the nation's lowest recycling rates.

They were only a few months into the campaign, said Gillham, a vivacious sixtysomething with tight curly hair and a gap-toothed smile. She got involved through a friend in Houston who had long served as an unofficial river keeper of the bayous that cut through Houston. He'd become convinced that the only way to stop the pileup of riverine litter was through a bottle bill, and he'd asked Gillham to help run the crusade. "He knew I had a little green heart," she explained. She, in turn, recruited her friend Wood. The two soon decided to quit their jobs to work full-time on developing the bottle bill.

Wood, also in her early sixties, is new to political activism. Though a longtime supporter of the Sierra Club, active in local prairie restoration and animal rescue programs, she'd never before done outright campaigning. But a bottle bill seemed such a straightforward no-brainer solution to the problem of plastic waste, she just had to take it on.

Though the specifics of deposit laws vary state to state, the mechanics are largely the same: A store selling a bottle of Coca-Cola, for instance, pays a deposit on that bottle to the Coca-Cola distribu-tor. When I buy that Coca-Cola in a bottle-bill state, I pay the store a deposit (usually a nickel or a dime). When I'm done, I take the empty back, and my deposit is refunded. The store then recoups its deposit on that bottle from the distributor, plus a handling fee, gener-ally ranging from one to three cents, to cover the cost of dealing with the empty bottle. In some states, the empties are collected by stores, in others at redemption centers or in reverse-vending machines.

These laws work because they make sense, said Wood. "Bottle bills have proven where there's a financial incentive, people will recycle," she said, and she began running through the statistics: bottle-bill states have at least twice the recovery rates as non-bottle-bill states. Michigan, where bottles can be returned for ten cents apiece—the highest amount in the country—gets more than 90 percent of PET bottles back. It also has one of the country's lowest litter rates. You put value on that discarded empty, and someone will pick it up and return it, she argued. Otherwise, it stands little chance of getting into the recycling stream.

I didn't have to go any farther than the lobby of the hotel hosting the conference to see her point. Sitting in the coffee shop there, I watched one of the conference participants diligently hand his empty juice bottle to the barista, clearly assuming she would pop it in a re-cycling bin. She took it with a smile, waited until he'd walked out of view, then tossed the bottle in the trash. Even though there was a drop-off recycling center a block from the hotel, the shop wasn't sav-ing empties. In Houston, near Wood's hometown, curbside recycling is hit or miss, depending on the neighborhood, and the program is so underfunded that there's a waitlist for recycling bins. (On learning about the waitlist, a San Francisco activist held a fundraiser to send the city two hundred bins.) There's no curbside recycling at all in the suburb where Wood lives, so she and her husband drive fifteen miles to deliver their empty soda bottles to Houston's city recycling center. "Not too many people are going to want to do that," she said.

To gather support for the bill, she and Gillham are trying to pre-sent it as not only an environmental plus, but an economic one—a

generator of jobs and money for the state. All day they'd been working the exhibit hall, hoping to drum up support from the industry.

They'd come to the right place. A running theme at the conference
was reclaimers' complaints about the difficulty of getting a clean and
reliable supply of PET bottles. They depend heavily on the bottle-
bill states. Indeed, those states supply most of the plastic bottles that
are recycled in this country, especially those bought by processors
who are recycling used bottles into new ones. But the domestic supply isn't enough, as I learned from reclaimers at the conference. So
many bottles are sold to China that American reclaimers commonly
buy used bottles from Mexico, Latin America, and Europe to keep
their plants operating. The market for clean bottles "is so ripe," said
Gillham, "all we have to do is collect this stuff and they're ready to
buy it up."

Still, the two women know they are in for an intense political fight.
Aside from Hawaii, which enacted a bottle-bill law in 2002, no state
has succeeded in passing a bottle bill since 1986, and there's an ongoing push to repeal existing laws. The fiercest opposition has come
from the beverage industry, which would be on the hook for the cost
of collecting empties from stores and redemption centers and disposing of them. The industry complains that it is being singled out from
other producers and that the cost of the deposit could deter sales
(though there's no evidence that beverage sales are any lower in states
with bottle bills). Grocers too have fought bottle bills, arguing they
don't want to be responsible for the hassle of dealing with and stor-
ing empties. Interestingly, the plastics industry has been mostly silent
on the bottle-bill debate: it hasn't fought proposed bills outright, but
neither has it supported them.

Indeed, unlike the paper, steel, and aluminum industries, the plastics industry historically has done little to support recycling, except
when it's under political pressure, as in the current fight over plastic bags. Howard Rappaport, a longtime consultant to the plastics
industry, explained why. The only players with significant financial
resources to invest in recycling are the resin producers, the major
oil and chemical companies, he said. But their top priority is "to

make and sell virgin plastics." As long as oil and gas prices are reasonably stable, there's no financial incentive for the Dows, DuPonts, and ExxonMobils to get into the recycling business. Nor do they want to alienate the beverage companies that buy their raw plastics to make bottles. Meanwhile, the companies that make plastic products — which might be expected to have an interest in using recycled materials — are too fragmented a constituency to put together an all-out campaign for more recycling, said Rappaport. "The guy making trash bags has nothing to do with the guy making bottles. He's got nothing to do with the guy making toys. It's so fractured that nobody can get enough critical mass and money together" to put into developing the recycling infrastructure.

The aluminum, steel, and paper industries are vertically integrated, which has made it easier and financially more appealing for them to incorporate recycling. So no surprise: those materials are recycled at three to eight times the rate of plastics. The sprawling assemblage of networks that makes up the plastics industry, said Rappaport, "wasn't built to incorporate recycling."

Whether it's recycled or thrown away, an empty Coke bottle is just the tip of a giant waste iceberg that starts much farther upstream. In waste circles, people like to point out that for every pound of trash put out at the curb, another seventy pounds is generated in the manufacture and production of their source materials.

To really reduce waste, says the EPA, you need to prevent it from being created in the first place. It sounds so obvious! Yet in our eagerness to relieve pressure on landfills, we've put the main focus on end-of-life solutions — managing the stuff already identified as waste — rather than reducing its production at the source. The mantra of solid-waste management has long been reduce, reuse, recycle. But the fact is most of the efforts have been centered on the last. And as a result, we haven't altered that essentially one-way flow of materials from factory to garbage can.

"The recycling movement has missed the forest for the trees," complained Bill Sheehan, a longtime activist in the recycling and zero-waste movements. "The amount of materials flowing through the economy has gone up and up and up."

The evidence is clear in the reports the EPA assembles every year documenting the country's garbage habits. In 1970, the average American generated 3.25 pounds of trash each day and sent 3 pounds to landfill. In 2008, the average citizen created 4.5 pounds of trash a day and sent 2.4 pounds to landfill. We're recycling more, but nowhere near enough to offset our growing consumption. Forty years after the first recycling programs got under way, Americans are producing more plastic packaging than ever and throwing away nearly as much. That fact has led some critics to complain that recycling is an empty exercise, little more than "a rite of atonement for the sin of excess," as *New York Times* journalist John Tierney put it in a famous anti-recycling tirade published in 1996.

But Sheehan argues that recycling can be made to work. We just need to shift the burden from consumers to producers. He's part of a movement pushing for a set of policies that fall under the loose rubric of extended producer responsibility. The basic concept of EPR is simple: make the producer responsible for a product's entire life, not just while it is in use but also after it's been used. As one EPR expert succinctly explained it: "you make it, you deal with it."

Sheehan maintains this is perfectly logical if you look at how the waste stream has changed over the past hundred years. When municipal waste systems were established, most of the trash collected consisted of food scraps and ash (from stoves and fireplaces). Just a tiny fraction was manufactured items, such as paper and rags and bottles. Today, more than three-fourths of municipal waste consists of manufactured products such as PET bottles. Having taxpayers cover the costs of disposing of the tens of millions of pounds of manufactured goods is a giant unfunded mandate, said Sheehan. "It's enabling producers, in fact *encouraging* producers, to make throwaway products."

Sheehan and other EPR advocates say it's time for private industry to start paying for the collection, disposal, and recycling of those products. Making producers financially responsible for end-of-life waste management gives them the incentive to be less wasteful from the get-go, Sheehan explained. If they have to pay to recycle their products, they'll be more likely to design them for better recyclability, picking materials, such as PET, that are compatible with existing recycling streams. Such schemes create a positive feedback loop that municipal recycling simply cannot.

Some of this is back to the future. Those refillable-bottle programs that disappeared with the arrival of PET bottles? They were an early form of EPR. Bottle bills also represent a version of producer responsibility, since they require the industry to take the bottles back. But producer responsibility was launched as a movement in 1991 when Germany introduced the first explicit EPR law, the Teutonicly titled Ordinance on the Avoidance of Packaging Waste. It required manufacturers and brand owners to provide for the recovery and recycling of the packaging associated with their products. At the time, German landfills were nearing capacity — packaging waste was a major culprit — and the country was starting to have to export its trash to neighboring countries. The government set ambitious goals, requiring that 64 to 72 percent of packaging materials be recovered for recycling, for example, but left it to industry to determine how to fulfill them.

The result was the Duales System Deutschland and its Green Dot program, which has become a model for managing packaging waste around the world. Companies pay licensing fees to a nonprofit corporation for the right to put a logo — the famous Green Dot — on their packaging; this tells consumers they can put those packages in the recycling bins. The licensing fees help cover the costs of the for-profit corporation that then collects, separates, recycles, and processes the green-dotted waste. Over 75 percent of all packaging in Germany now carries a green dot, including, of course, PET bottles. And the program has expanded across Europe; tens of millions of packages sold in European stores now bear the Green Dot logo, which, unlike our resin code system, is a genuine guarantee that they will be recycled.

The law has accomplished most of its goals — and more. It has boosted recovery rates of packaging waste to more than 76 percent and hiked up plastics recycling to 60 percent; in some years, the numbers are even higher. It has also stimulated the development of new sorting and recycling technology to help the country meet those mandates. Germany's law became the model for a similar measure passed by the European Parliament in 1994. While that law set recycling targets, it left it to individual countries to determine how they would go about meeting them. Different countries have chosen different routes — varying arrangements of public and private initiatives — but the overall story is much the same. Less packaging is going to landfills, even while consumption has continued to grow. Overall, Europe now diverts more than half of its postconsumer-plastic waste from landfills and is recycling 29 percent of its plastic packaging. This doesn't sound that impressive until you consider that it's an average figure that includes the recycling rates of countries with still dismal records, such as Greece, which recovers scarcely 10 percent of its waste, in addition to places such as Germany and Denmark, which are closing in on 100 percent recovery of packaging waste. Sweden recycles 80 percent of its PET bottles, thanks to the equivalent of a national bottle bill.

It's worth noting that a large portion of the used plastic Europe recovers is burned to make heat or electricity, a technology known as waste-to-energy. There are about four hundred waste-to-energy plants scattered across the continent, and the EU counts what they do as recycling, which is why many European countries boast such enviably high recycling rates. Because landfill space is in such critically short supply, environmental concerns about the plants have gained little political traction. In the United States — where there's room to continue the debate, so to speak — there are only 87 waste-to-energy plants, most on the congested East Coast, and no new ones have been built since the mid-1990s. Yet as plastic packaging increasingly comes under fire, leaders in the plastics industry have begun advocating for such technology as a way to deal with plastic waste.

In addition to spurring recycling efforts, the EPR laws have galvanized manufacturers and brand owners to make significant changes in their packages. They are using less material and eliminating unnecessary packaging—no more toothpaste tubes in cardboard boxes, fewer packages made of rarely recycled materials, such as PVC, and more refillables and concentrates. There's also much wider use of refillable bottles—glass as well as PET—which studies have shown have the lightest environmental impact. The Green Dot law cut packaging consumption in Germany by 7 percent in its first four years, during which time, it's been estimated, U.S. packaging consumption increased by 13 percent.

The idea of EPR is spreading. More than thirty countries have packaging take-back laws, and many have followed Europe's example by passing laws that require producer responsibility for hazardous materials, such as mercury, and for some products, such as electronics and cars. The United States has long been a holdout on demanding producer responsibility, but that is starting to change. At least twenty-four states have enacted producer-responsibility laws covering electronic waste, and places as diverse as Florida, Minnesota, Indiana, and Maine are developing laws requiring producers to step up and take responsibility for the disposal of products such as unused pharmaceuticals, mercury thermometers, thermostats, and lamps. In California, Vermont, Oregon, and Rhode Island, the EPR concept is being embraced as a general principle guiding waste policy. The organization Sheehan heads, the Product Policy Institute, has established several product-stewardship councils—groups of policymakers, activists, and businesspeople who work on promoting EPR policies and laws—in California, Oregon, Washington, Vermont, New York, and Texas (as well as British Columbia). "Not only is it sweeping state legislatures," he said. "Industry is taking this very seriously."

Even without the explicit prod of EPR legislation, a growing number of companies are looking for ways to reduce their packaging footprints. What's good for the planet, many have realized, is also good for the bottom line. One forum for such efforts is the Sustainable Packaging Coalition, an organization created by architect William McDonough to implement some of the ideas laid out in his 2002 manifesto for industrial change *Cradle to Cradle*. In that book, McDonough and coauthor Michael Braungart argued that waste is a sign of fundamental unsustainability. Nature, they pointed out, "operates according to a system of nutrients and metabolism in which there is no such thing as waste." They used the example of a cherry tree. The tree produces blossoms and fruit, and though many fall to the ground, they are not useless there. They provide food for birds and insects and microorganisms, and they enrich the soil. In natural systems, "Waste equals food. This cyclical, cradle-to-cradle biological system has nourished a planet of thriving, diverse abundance for millions of years." Human manufacturing, they argued, should operate the same way, so that all the things we make and the byproducts that are generated can be cycled back into either the metabolism of industry or nature.

That principle is the motivating spirit of the Sustainable Packaging Coalition. The group comprises more than a hundred of the world's leading materials manufacturers, packaging companies, consumer products companies, retailers, and waste haulers, including Dow, Coca-Cola, Pepsi, Apple, Dell, Starbucks, Procter and Gamble, and, of course, Walmart, which has shaken the global supply chain with its measures to dramatically reduce packaging waste. The group meets quarterly to share ideas, strategies, and the challenges of designing cradle-to-cradle packaging.

I attended one of its meetings, and after observing the official presentations and the hallway conversations, I realized just how complicated its goal can be. For starters, consider packaging's many purposes: keeping food fresh, protecting contents from damage, deterring theft, and, of course, enticing shoppers, among other things. Any of those factors can be affected when you start tinkering with the way a product is packaged. "I can take all the packaging off food and within a month I'll see food waste double" because the food will go bad, said Helen Roberts of Britain's big-box store chain Marks and Spencer in one session. Or, as a representative of Procter and Gamble observed in another session, consumers may not understand or share your goals. The company recognizes that small bottles of concentrate are environmentally preferable but they are a tough sell because shoppers still have a "primal" sense that "bigger is better." In another session, someone from Earthbound Farms, the country's biggest supplier of organic greens, described the company's efforts to use packaging that is as earth-friendly as its products. The company had taken pains to make sure that the clamshell containers used to protect delicate produce that was being transported around the country were made of PET, a recyclable plastic. Still the Earthbound Farms rep looked dismayed when a woman attendee — who was clearly not a packaging professional — stood up in the back, radiating anger. I want you to know, she announced, "I've made a personal decision not to buy your lettuce because it's in plastic!"

Even the environmental imperatives of sustainability can be difficult to reconcile. Take the example of the juice boxes and soup boxes, which began appearing on store shelves in the 1990s. From an energy perspective, the packages are very efficient. They use less material and are more lightweight than cans or even plastic bottles. They take up less space in boxes and pallets and last longer on the shelf. Those kinds of efficiencies fit squarely within several of the Sustainable Packaging Coalition criteria for a sustainable package. But juice boxes are such complicated sandwiches of plastic, metal, and paper that they cannot be recycled, which clashes with another criterion of sustainability. In today's world, there are no perfect choices; all we can do is be aware of the tradeoffs. As SPC staffer Anne Bedarf noted, "There's no such thing as a sustainable material or package." There's only an ongoing journey to improved sustainability.

The empty twenty-ounce Diet Coke bottle sitting on my desk is, in many ways, a reflection of that journey and its challenges. Coca-Cola has been concerned with the environmental impact of its packaging for decades, said Scott Vitters, the company's sustainable-packaging guru. Vitters, who has worked for Coca-Cola since 1997, is a resource wonk, prone to talking in almost impenetrable sustainability-speak: "Our objective is how do we maximize the contribution of packaging to economic, environmental, and social sustainability through preventing and eliminating waste across the entire product value chain." Translation: packaging waste costs the company money and goodwill. So Coca-Cola looks for ways to reduce the amount of materials that go into its packages and then tries to ensure that its packages aren't wasted. To that end, said Vitters, it's been a longtime adherent of the three Rs: reduce, reuse, and recycle. Indeed, it has made a public commitment to achieving zero waste.

Compared to the PET bottle that Nathaniel Wyeth introduced, the bottles Coca-Cola and other beverage makers use today are significantly lightened and streamlined. The family-sized bottles have lost the flat-bottomed high-density polyethylene base that marked the original ones; it was a pain for recyclers. The bottle walls have been thinned, allowing them to be made with a third less plastic. The labels are now smaller and affixed with adhesive that is easier to remove, to improve recyclability. And even the caps have been shortened, decreasing the amount of plastic needed by 5 percent, for a savings of forty million pounds of plastic a year. "It adds up quickly," said Vitters. That twenty-ounce Diet Coke bottle I've got in front of me is the lightest PET soda bottle on the market. Other packages, such as its Dasani water bottles, have also been light-weighted, reducing the amount of plastic used by a hundred million pounds a year.

Indeed, the Dasani bottle is one of Vitters's proudest achievements. Coca-Cola designers were planning to launch Dasani in a dark blue bottle until he pointed out that the dark tint would make the bottle unrecyclable, since PET reclaimers need colorless material. They decided instead on a much lighter shade. The caps were also made to be more easily recycled. "Every aspect of that package was designed for end of life," he said. Coca-Cola Japan accomplished a similar feat with a super-lightweight bottle for its recently introduced line of bottled water I LOHAS (for Lifestyles of Health and Sustainability): a container so minimal it can be crushed into a twelve-gram wad of plastic, nearly half the weight of any other PET water bottle. Of course, it's not quite clear how the product inside that feather-light bottle fits into a true Lifestyle of Health and Sustainability, given the energy that goes into bottling and selling a substance that most consumers can enjoy from their home faucet for a fraction of the cost. (And let's not even get started on the health and environmental sustainability of Coke's main product.) Such are the paradoxes of sustainable packaging.

Coca-Cola pioneered the use of bottles with recycled content in 1991, but it's followed through with more consistency in other countries, especially those with strict EPR laws. Globally, the company has sought to ensure that the resin used in its bottles contains at least 10 percent recycled content, but in many places the recycled content is as high as 25 or even 50 percent. Despite promises the company made in the early 1990s, the recycled content of bottles sold in the United States remains, at most, 5 to 10 percent. (Those rates apply to the resin used to make the bottles, not the amount of recycled content contained in each bottle.) Vitters said one reason U.S. bottles have lower amounts of recycled content is the difficulty of collecting used bottles in this country (the very problem that could be addressed by more deposit laws).

The company recently adopted the goal of recovering or collecting 100 percent of all the bottles and cans it sells in the United States. That means continuing to develop new uses for used bottles, such as the T-shirts and chairs it now makes from old Coke containers. But more important, it means expanding its investment in the recycling infrastructure, most significantly by plunking down forty-five million dollars in 2007 to build the world's biggest bottle-to-bottle recycling plant in Spartanburg, South Carolina. (The company already operates similar closed-loop recycling plants in Austria, Mexico, Switzerland, and the Philippines.) The South Carolina factory has the capacity to recycle a hundred million pounds of PET a year — the equivalent of two billion twenty-ounce Coca-Cola bottles. Once the plant is fully online, Coke hopes to put 25 percent recycled content into the bottles sold in the United States. (Coke could raise the recycled content even higher, but not to much more than 50 percent; experts say that beyond that level, the resin starts to degrade and darken.)

So is this the real thing? Some skeptics question Coke's commitment, since for years it has bitterly fought the one type of producer-responsibility measure that's been shown to increase recycling and reduce plastic waste: bottle bills. The deep pockets of Coca-Cola and other major beverage makers is the main reason we have so few deposit laws. Vitters insisted the company wasn't opposed to EPR legislation, just to laws that focus only on bottles and cans. "We need to focus on all packaging and printed materials. And then you hold industry accountable for the costs of recycling," he said, mentioning with approval the European EPR systems for packaging. Vitters makes a reasonable point. Yet if that's really the company's position, why hasn't it worked proactively with EPR advocates to develop legislation instead of fighting to crush every proposed bottle bill?

In early 2010 the company unveiled its latest step toward sustainability: the PlantBottle. It's a PET bottle, but the feedstock for part of the plastic comes from plants rather than oil or natural gas. One component of the PET, monoethylene glycol, is derived from Brazilian sugar cane, while another, terephthalic acid, is still processed from fossil fuels. "About a third of the bottle is now plant-based, but we're working hard to get it up to a hundred percent," said Vitters. But even though the PlantBottle comes from new types of feedstock, it fits right into the existing PET recycling stream. To Vitters, this bottle is the ultimate expression of cradle-to-cradle thinking, the embodiment of the company's commitment to zero waste. As he put it, "The PlantBottle ties this all up."

Can plant-based plastics like the new Coke bottle really help deliver a cradle-to-cradle future, a future where empty bottles, like scattered cherry blossoms, are a valued nutrient of production rather than a devalued waste? It's a comforting vision, and in our long, turbulent relationship with plastics, we might be forgiven for seeking comfort where we can find it. Clearly, bioplastics is where the industry as a whole is headed. But as anyone who has ever been in couples' counseling knows, envisioning a healthier future is one thing. Putting new insights into practice is where the heavy lifting begins.

The Meaning of Green

ONE EVENING IN 1951, so the story goes, a New York businessman named Frank McNamara was out dining with friends. The check arrived, and McNamara was chagrined to realize that he had left his wallet at home. That chastening experience led him to invent the Diners Club card, which allowed members to charge meals at participating restaurants and pay their tabs at the end of the month. Apparently he wasn't the only one who'd suffered the embarrassment of finding himself cashless: within a year, twenty thousand people had signed up for Diners Club cards. The card itself was nothing remarkable, just a wallet-sized square of cardboard, but "the idea behind it — a third party facilitating a 'buy-now, pay-later' process — was revolutionary," as one history of credit cards noted. The concept of money had become attenuated, the line between cash and credit blurred. Money's new identity was lodged in a symbolic card that allowed you to pay even when you couldn't flash the legal tender. Money had taken on a new kind of plasticity.

Money would be truly yoked with plastic a few years later, in 1958, when American Express introduced the first plastic credit card — yet another in the tide of transformative plastic goods that Americans began to embrace in the mid-twentieth century. AmEx promoted the plastic card as a step up from the flimsy paper ones then in use, promising it would "better withstand day-to-day use." Implicit here was the notion that this card wasn't a convenience for the occasional dinner out but a tool for daily life. We would no longer be bound by bank hours or the approval of a loan officer. We could buy what we wanted — any time, any day — and pay later. Unlike the prior card issuers, American Express, MasterCard, and others started to provide revolving lines of credit in the mid-1960s, which allowed customers to carry balances from month to month. Credit wasn't a new concept, of course, but its instant availability was radical. Now we were fully released from the constraints of tangible money, our purchasing habits no longer limited by "cash on hand." We were free to consume, whether we could afford it or not.

It's no wonder that within another decade or so, credit cards were so commonplace the very word *plastic* became synonymous with money, edging out phrases that evoked the texture of tangible cash, the metallic clinking of *two bits,* the sandpapery feel of *sawbucks.* Any reader knew just what novelist Dan Jenkins meant when he wrote, "She had a whole purse full of plastic," in his 1975 novel *Dead Solid Perfect,* the first recorded use.

Today, those plastic cards are the chief currency of commerce. Or as the website of one card manufacturer stated grandiosely, "A plastic card is a physical device that links people to civilization." Three-fourths of American adults have at least one credit card; most have three or more. But the credit card isn't the only plastic taking the place of cash in the average wallet. Four out of five Americans own debit cards, and one in six has a prepaid card to buy gas, make phone calls, or use for general purchases. Plastic cards are also increasingly the stand-ins for gifts, especially for giftees one doesn't know well — the doorman, a coworker, a distant relative. Miss Manners may complain that the impersonality of gift cards has "taken the heart and soul" out of giving, yet they offer such a convenient and stress-free mode of appreciation that ten billion are now created annually. Who knew we were so altruistic?

Over the course of their evolution, these small vinyl rectangles have become a canvas of marketing goals and cultural preoccupations. Card issuers have played on status consciousness with color-coded luxury cards, from American Express's first gold card to Visa's popular Austin Powers titanium card, which was promoted with the slogan It's Titanium, Baby! In the 2009 movie *Up in the Air,* the object of desire for the protagonist played by George Clooney was the elusive carbon-black million-mile frequent-flier card. Banks have appealed to emotional connections with affinity cards, first popularized by Visa in 1989, when it cobranded a card with the National Football League. Today, chances are good that I can find an affinity card for whatever cause I care about, from the National Rifle Association to People for the Ethical Treatment of Animals. Or at the very least I can choose an image for the front of the card that conveys what's dear to my heart, whether it's puppies, my alma mater, my favorite band, or my family.

It's not only the terms of the card or the adorning imagery that offer a key to the Zeitgeist. It's the material of the card itself—that five-gram sheet of plastic. At a time of growing concern about plastic's toll on the environment, cards are undergoing a total transformation. These facilitators of consumption are going green.

I'm holding in my hand my new Discover card. It looks like any regular plastic credit card, yet it's made out of a kind of polyvinyl chloride that I'm told will harmlessly biodegrade when I throw it away. The back of the card is colored an earthy brown and bears the word *biodegradable.* When I ordered the card over the phone, I was told I could pick from several options for the front: plain gray, an American flag, a polar bear, a panda bear, mountain scenery, a beach. None seemed especially relevant to the problems of plastic pollution, but with thoughts of the Pacific vortex in mind, I chose the beach, not realizing the image I'd wind up with would have the supersaturated colors and unreal look of a Club Med brochure. "Well, it is a *Discover* card," my husband said when I showed it to him. "They want you to spend money to discover the world."

Discover introduced these new "environmentally friendly" cards in late 2008, just in time for the Christmas season. "We hope this will appeal to those interested in living a greener life," a company spokeswoman said at the time. The company won't say how many green-living customers have taken them up on the option, only that "We are encouraged by the results thus far which have exceeded expectations."

PVC, you'll recall, is the plastic environmentalists hate more than any other, the one known in Greenpeace circles as "the poison plastic." Nearly all credit cards, as well as gift cards and debit cards, are made of PVC and have been since the American Express debut. Card manufacturers like PVC because it's easily processed, offers the right blend of rigidity and flexibility, and is durable enough to last the standard three-to-five-year term of a credit card.

Granted, environmental issues rank well below debt issues when it comes to the hazards usually associated with credit cards. Nor are credit cards the products activists usually point to when warning about the dangers of PVC. Yet in Plasticville, even small objects like credit cards add up. By one estimate, there are more than 1.5 billion credit cards in use in the United States. A stack of them all would reach more than seventy miles into space, the *New York Times* calculated; it would tower nearly as high as thirteen Mount Everests placed one on top of the other. But the natural laws of erosion and degradation that whittle away mountains would scarcely dent that polymer peak. Even a single PVC card can persist for decades, if not centuries, and each year, we toss away more than seventy-five million. And that's just credit cards; those tallies don't include the much greater number of gift cards, prepaid cards, hotel keycards, and other varieties of plastic used to transact life these days.

The thought of all those plastic cards accumulating in landfills was what motivated the man who is responsible for the plastic used in Discover's allegedly ecofriendly card. For twenty years, Nevada businessman Paul Kappus had sold PVC to the makers of credit cards, and he'd often thought there should be a way to make the used cards decompose. He spent several years talking to scientists and chemists, searching for some chemical that might make the PVC mortal. "I tried all sorts of different things, like the enzymes that eat cat urine off the floor. You wouldn't believe what I tried. All of it was a gross failure." Until one day he stumbled across what he says was an obscure technology that turned out to work.

Kappus was vague on the details, saying his formulation, BioPVC, is a trade secret. He'll state only that it involves a special additive blended into the PVC that acts like bait to the microorganisms that are ubiquitous in the environment, including in landfills. The additive doesn't affect the card's durability while it's in use; it'll stand up to years of swiping and stowage in a wallet. But deposit that card in a landfill or compost pile or any similarly "fertile environment," and, according to Kappus, it will draw hordes of microscopic critters that can take it apart. "They actually eat it, believe it or not," he said. Even a card that's litter on the ground will be scavenged, he claimed, without leaving behind any of the polymer's toxic precursor vinyl chloride. He said the card would be fully degraded within ten years, a blink of an eye compared to a regular PVC card.

This all sounded really wonderful—until I started talking to experts on biodegradability.

"That's a load of hooey" was the reaction of Tim Greiner, a Massachusetts sustainability consultant. Like other experts, he was dubious that PVC could be made to harmlessly melt away. But even if it did work, Greiner questioned the need for it. Biodegradability is a nice solution for litter, perhaps. But credit cards aren't generally littered. So, Greiner asked, "What is the problem this card solves?"

What problem, indeed?

It was a useful question to bear in mind as I started wading into the thicket of "green" plastics. What I found was a broad and sometimes bewildering variety of products made with or packaged in resins that manufacturers claim are safer for the environment and our health, including chip bags, water bottles, cell phones, BB gun pellets, diapers, carpets, cutlery, ballpoint pens, socks, cosmetic cases, plant pots, Easter-basket grass, flip-flops, and trash bags "with a conscience." Some, like the Discover card, involve conventional plastics with a green twist. Others are made from alternative "biobased" polymers: for example, the Apple iTunes gift card my daughter recently got for her birthday is made of a corn-based plastic.

Green plastic might sound like an oxymoron, but it's one of the industry's fastest-growing fields. Production of biobased polymers has been expanding at the rate of 8 to 10 percent a year and is expected to grow much faster in coming years. There's so much excitement about bioplastics that it's tempting to describe their rise as a boom. But when I used that term with Ramani Narayan, one of the country's leading biopolymer experts, he reminded me that biobased plastics are still only a drop in the resin bucket, less than 1 percent of global plastics production. The field is in its infancy, with a steep technological learning curve ahead. Nonetheless, a recent study estimated that bioplastics could one day replace as much as 90 percent of today's plastics. Said Narayan, "This is the future of plastics."

There's no mystery why. A century into mankind's love affair with plastic, we're starting to recognize this is not a healthy relationship. Sure, plastics have been a good provider, but that beneficence comes with many costs that we never even considered in our initial infatuation. Plastics draw on finite fossil fuels. They persist in the environment. They're suffused with harmful chemicals. They're accumulating in landfills. They're not being adequately recycled. In short, they exemplify shortsighted thinking about the long-term impacts of manufactured materials and represent an unsustainable wasting of resources. Environmentalists have been making that case for years. Now even the plastics industry is coming to the same conclusion. As a Dow executive told *BusinessWeek,* "Our whole industry agrees that plastics have to be more sustainable."

In any event, it's not as if we can get a divorce. Plastics are one of the material foundations of modern life, and in many contexts, that's a good thing. We want our solar panels, bike helmets, pacemakers, bulletproof vests, fuel-efficient cars and airplanes, and, yes, even much of our plastic packaging. As they did in the late nineteenth century, plastics have a vital role to play in a world of dwindling natural resources. And that will be even truer in coming decades as we grapple with climate change. More and more of our decisions about how to build our homes, transport ourselves, and package our stuff will be driven by carbon calculations. By that measure, lightweight, energy-efficient plastics can offer extraordinary opportunities.

But to live in harmony with plastics, we have to change the terms of the relationship. We need to develop plastics that are safer for people and the planet, and we need to deploy them more responsibly. And that means change on the parts of all residents of Plasticville: the producers of plastic things, such as credit cards, and the consumers who use them.

What constitutes a green plastic? Though there's plenty of debate, most would agree that one starting place is the use of renewable raw materials, a quest that, ironically, takes the industry full circle, back to plastic's earliest roots as a material derived from plants. Remember celluloid? That wasn't the only plant-based polymer. Throughout the early decades of the twentieth century, there was widespread interest in making other types of plastics from agricultural crops, such as corn or legumes or soybeans. Indeed, agricultural interests competed fiercely with the nascent petrochemical industry to capture the market on polymers.

Henry Ford, who was eager to find industrial uses for surplus crops, put his money on soybeans. He often claimed it would be possible to grow most of an automobile. To that end, he planted thousands of acres of soybeans and converted one of his plants at River Rouge to the production of a soy-based plastic. The typical 1936 Ford had ten to fifteen pounds of soy plastic in its steering wheel, gearshift knob, window frame, and other parts. In 1940, Ford famously invited reporters to see a "farm-grown" car. Ever the showman, the septuagenarian hefted an ax and swung it hard against the back of the custom-built car. Instead of crumpling, the panels bounced back into shape. Or, as a reporter for *Time* magazine put it, "the fenders of the Buck Rogers material . . . withdraw from collisions . . . like unhurried rubber balls."

But Ford didn't get a chance to make more than one plastic car before World War II broke out. Then even he couldn't buck the advantages petrochemistry had to offer. Oil was inexpensive and plentiful, and unlike soybeans and other crops, it didn't depend on growing seasons. What's more, the plastics made from fossil fuels were more waterproof and versatile than those made from soy. Only a few plant-based polymers survived that contest with petrochemistry, among them cellophane, celluloid, and the textiles rayon and viscose.

Now, with the era of cheap fossil fuels coming to an end, the petro-based resin manufacturers are combing the natural world for new building blocks. They're looking at agricultural crops, such as corn, sugar cane, sugar beets, rice, and potatoes. And they're exploring historically undomesticated sources of carbon, such as switchgrass, trees, and algae. "Carbon is carbon. It doesn't matter if it was sequestered in an oilfield 100 million years ago or six months ago in an Iowa cornfield," as one bioplastics executive put it. As it turns out, that's not entirely true.

Much of the current effort amounts to pouring new feedstocks into old bottles, using plants to make conventional polymers of all sorts. Brazil's petrochemical giant Braskem, for instance, is building a 450-million-pound-capacity plant to make sugar-cane-based polyethylene, the same material Coca-Cola is deploying in its PlantBottle. (To publicize the venture, Braskem teamed up with a toy maker to produce the game pieces for a Brazilian version of Monopoly, which they called Sustainable Monopoly—perhaps appealing to the capitalist's inner environmentalist, or vice versa.) Sugar-cane ethanol already fuels most cars in Brazil, and the country is hoping its vast cane plantations will allow it to become a global hub for cane-based plastics as well. Dow Chemical has grabbed the hook, teaming up with local ethanol producer Crystalsev to build its own cane-based-plastics plant in Brazil.

Likewise, various resin companies have worked out ways to make polypropylene and PET from sugar, nylon from beets, and polyurethane from soy. That last has given Ford a chance to fulfill its founder's dream: the company has used soy-based polyurethane cushions and padding in more than 1.5 million cars and eventually plans to make all the plastic parts in its cars from various types of composta-ble plant-based plastics. Meanwhile, Solvay, one of the world's largest producers of vinyl, is exploring plant sources for the ethylene used in making PVC.

These bioplastics may or may not be an improvement on their fos-sil-fuel-based relatives. Certainly they have a far lower carbon foot-print, by virtue of their renewable feedstocks. Some are, in principle, recyclable, and many are compostable. Yet there's no guarantee they are manufactured with less harmful chemicals or contain any fewer worrisome additives. A credit card made of a plant-based PVC, for instance, would still contain toxic vinyl chloride. "Just because it's biobased doesn't make it green," pointed out Mark Rossi, a Boston-based activist and researcher who has spent much of his career think-ing about how we can forge a healthier relationship with plastics. For years that was a fairly lonely pursuit. But lately, he's found he has a lot of company. "I was on the plane coming back from Chicago or Detroit or somewhere," he recalled when we met for coffee. "I was talking to this woman I was seated next to who was a mom. And she was totally up on all this stuff, like bisphenol A–free bottles. Five years ago it wasn't on most people's radar."

Rossi got interested in polymers in the late 1980s, during the hue and cry over the country's solid-waste crisis. That was back when McDonald's was under attack for selling their burgers in Styrofoam clamshells, he recalled, and the company proposed putting "little Mc-incinerators" in their stores to get rid of the waste. "This was their solution," he said with a slight chuckle, still struck, after all these years, by its absurdity. The controversy prompted the subject of his master's thesis in environmental science: a life-cycle assessment of polystyrene packaging. He went on to work on an influential 1992 study that to everyone's surprise — including his own — showed that plastic packag-ing wasn't the thorough environmental villain often supposed. The study found that some of the most significant environmental impacts of packaging are a consequence of weight. Plastics permit smaller and lighter packages, which require less energy and resources to produce than glass or paper or other materials. From there Rossi moved on to examining the problems of chemical additives, eventually working with Health Care Without Harm on their campaign to get PVC out of hospitals. Now he's research director of Clean Production Action, a group that works with businesses and government to push for the use of safe chemicals and sustainable materials in manufacturing.

Those efforts have taught Rossi a few things. One is that "all plastics aren't created equal." Some are greener, or have the capacity to be greener, than others. That's as true of biopolymers as it is of petrobased plastics. As Rossi learned, there are decisions made at every stage of a plastic's life cycle that determine its health and environmental impacts — from choice of feedstock and how it is processed or grown, to the ways in which the polymer is manufactured and processed, to how the product is used, to the available options at the end of its useful life.

With that framework in mind, Rossi has helped develop two different scorecards for assessing plastics — one aimed at plastics in general, and another specifically for biobased polymers. Still works in progress, the scorecards are designed to help manufacturers, buyers for stores, and government agencies evaluate the environmental qualities of the plastic resins and products they are buying. A beverage company, for instance, could use the scorecard when weighing whether to bottle its water in a corn-based plastic or a polyethylene made from sugar cane. Or a buyer for a big-box store could use the scorecards to decide between a product that's packaged in polypropylene and one in PVC.

Each of the scorecards drills down into the nitty-gritty of plastics production, addressing questions such as: What crops were used in a bioplastic and how were they raised? What types of catalysts were employed to create a given polymer? What additives does it contain? What chemicals might be released in recycling? Is it being deployed in single-use products? How much recycled content does the plastic contain? The scorecards aren't life-cycle analyses — those tend to focus on the energy expended in making a product and aren't very effective at assessing issues such as the chemical impact. Rather, they provide an exercise in what Rossi calls "life-cycle thinking."

Sometimes life-cycle thinking delivers surprising answers. For instance, a plastic doesn't have to be biobased to start looking green. It's possible to make polypropylene without a lot of hazardous chemicals, and a polypropylene package that also contains high levels of recycled content could actually rate higher than a plant-based plastic. Likewise, a plant-based polymer can fall in its rating if the crops used to make it were genetically modified or sprayed with pesticides, or if harmful chemicals were used in its production. And some plastics, notably PVC, are intrinsically problematic. Even a biobased PVC would likely flunk due to the chlorine in the polymer chain that generates troublesome ripple effects across the life of the plastic.

At this point, it's tough for any plastic to earn straight As, acknowledged Tim Greiner, who worked with Rossi on the basic-plastics scorecard. He and Rossi intentionally set the bar high to try to spur the design of better plastics and plastic products. They're calling on industry to start doing the kind of life-cycle thinking about plastics that wasn't done decades ago when these dazzling new materials first burst onto the scene. "We wanted to define the true north, the direction [the industry needs to go] and give people the compass," said Greiner.

Some might say true north lies in a cornfield in Blair, Nebraska, site of NatureWorks, the world's largest producer of a wholly new kind of plastic derived from plants. NatureWorks makes a corn-based polymer called polylactic acid, or PLA. Chances are you've already encountered this plastic via its trademarked name, Ingeo. It's now in thousands of packages and consumer goods, including the "corntainers" Walmart uses to package fruits and veggies, the floor mats in Toyota Priuses, the casings of computers made by Fujitsu and NEC, the bottles used by Newman's Own salad dressings, the bottles of several brands of bottled water, the soda cups distributed by KFC, as well as gift cards issued by a number of retailers, like my daughter's Apple iTunes card. Gift cards and credit cards are potentially a huge market, said NatureWorks spokesman Steve Davies, adding that the card market is being driven by the card issuers rather than NatureWorks. "Everyone wants to get away from PVC," he explained. Visa and MasterCard have approved the use of Ingeo in credit cards, he said, and now "it's up to banks to adopt it." (Making the change can be tricky, as snack giant Frito-Lay learned. The company proudly announced that it was using PLA for the packaging of its SunChips, but less than a year later it switched back to conventional plastic because consumers had complained that the new bags were too noisy. The crackling sounds the bags made were louder than "the cockpit of my jet," groused an air force pilot in a video blog headlined "Potato Chip Technology That Destroys Your Hearing." He ran sound tests and claimed the bags registered 95 decibels, which is in the same range as a backhoe or lawn mower. "You feel guilty about complaining, since they are doing a good thing for the environment," another consumer told the *Wall Street Journal*. "But you want to snack quietly and you don't want everyone in the house to know you are eating chips.")

The building block of PLA is lactic acid, a natural compound that NatureWorks derives from the sugar in cornstarch. You don't have to use corn to make lactic acid; any starchy plant will do, including sugar beets, wheat, rice, or potatoes. The world's other major producer of PLA, Purac, has its main plant in Thailand and uses tapioca or sugar cane for feedstock. But NatureWorks is partly owned by Cargill, the world's largest purveyor of corn, hence the company's first factory was built on corn. Literally. The plant sits amid Cargill-owned cornfields. But Davies insisted the company is agnostic about potential feedstocks and plans to use whatever crop makes the most sense when it builds plants in other parts of the world.

Whatever the source, the lactic acid is converted to a monomer called lactide. Those molecules are then strung together into the polymer PLA. PLA is a fairly versatile plastic, able to perform many of the same functions of existing petro-plastics and capable of being processed with the same types of equipment. Like PET, it can be molded into clear, stiff shapes that are great for packaging. Like polypropylene, it can be extruded into the nonwoven fabric used in diapers and wipes, and like nylon and polyester, it can be spun into fibers for carpet and clothing. Among its shortcomings, however, is a melting temperature low enough that PLA bottles have been known to deform when left too long in hot cars. And it can't hold in carbon dioxide the way PET can, making it ill suited for the vast soda bottle market.

PLA was one of the first biopolymers to reach the market. But others are on their way, all using different technologies to get there. One of the most intriguing developments is the use of microorganisms to produce entirely new types of biopolymers. DuPont is employing *E. coli* bacteria to produce a biobased synthetic textile marketed under the brand name Sorona, and a Cambridge, Massachusetts, biotech company, Metabolix, has harnessed a different type of bacterium to create what it claims is a completely sustainable new biopolymer, which it calls Mirel.

Metabolix isn't using just any old bacteria. It's employing an unusual group of microbes that store energy not as fat but as a natural polymer called polyhydroxyalkanoate, or PHA. The microbes are ubiquitous, present in soil, air, the sea, even our bodies—and as a result, so is PHA. Metabolix founder Oliver Peoples learned about these microbes back in the late 1980s when he was a freshly minted chemist doing postdoc work at MIT. He saw in them a vehicle for producing a biobased polymer, a plastic that one day could even be grown in plants. It took more than a decade, but he eventually succeeded in genetically engineering those microbes to be "super producers" of PHA and devising methods of fermenting them in vats of dextrose corn sugar to produce huge quantities of the polymer. The microbes stuff themselves silly on the sugar and are now so efficient at converting it into PHA that it makes up 80 percent of their weight. The fluffy white polymer is then extracted, dried, and turned into pellets of Mirel. On a recent summer day when I visited Metabolix's offices in Cambridge, spokesman Brian Igoe walked me through labs where the Mirel was being grown. The room had a slightly yeasty smell from the yellowish microbial broth fermenting in stainless steel vats. Igoe likened the process to making beer. "But it's the most high-tech beer you've ever tasted."

The locked greenhouse down the hall, however, contained Peoples's ultimate dream. In that small rooftop space, a miniature prairie sprouted under banks of fluorescent lights. There were dozens of pots filled with switchgrass plants that had been genetically modified to grow PHA in their cells. "If you look at the stems and leaves of this switchgrass here, it has this kind of white finish to it," Igoe said, holding out a leaf for me to examine. That slightly white cast was the PHA; it made up about 6 percent of the plant's tissue. Nearby were several pots of tobacco plants, which were also growing PHA. (Tobacco DNA is so easily manipulated, the plant is the botanical equivalent of the lab rat in gene technology.) One day, the company hopes, plants such as these could be a source of useful plastic. The plant would be harvested and dried, the PHA extracted and turned into Mirel, while the leftover biomass could be burned for energy.

Peoples wasn't the first to see the possibilities of those micro plastics producers. Scientists at Imperial Chemical Industries in Britain tried to commercialize the technology in the 1980s with no success and passed the torch to Monsanto. In the late 1990s Monsanto reported that it had succeeded in creating a PHA plastic, and one credit card maker announced it would issue a "green earth" credit card from the new polymer. But before that could happen, Monsanto decided to pull the plug on its entire bioplastics division. Plant-based plastics were considerably more expensive than those made from fossil fuels, and Monsanto wasn't convinced there was a market willing to pay the added costs of going green.

But Peoples is certain the math can be made to work and that there is a market eagerly awaiting the arrival of green plastics. Grain giant Archer Daniels Midland has bought into Peoples's vision. In 2004, ADM formed a joint venture with Metabolix to start producing Mirel on a commercial scale. After repeated delays, the joint venture's plant in Clinton, Iowa, finally fired up in early 2010.

Bioplastics do cost more than petro-plastics — PLA is two to three times the price, though Davies said that price difference begins to evaporate when the price of oil passes eighty dollars a barrel. Mirel is even more expensive, but Peoples insists he isn't trying to compete with conventional plastics or, for that matter, PLA. He's positioning Mirel as "a premium product," a polymer that can be made into a film or foam or a rigid material and that has one big advantage over petro-based plastics: it readily biodegrades. That's the virtue Peoples is depending on to conquer just a handful of carefully picked end markets, including packaging, agricultural applications, and consumer products, such as gift cards. Indeed, in 2008 Metabolix struck a deal with Target to provide enough Mirel for millions of gift cards for the Christmas season that year. The cards gave Target an opportunity to shore up its green credentials while providing Metabolix a visible platform to announce its new plastic to the world.

Mirel and Ingeo are still a long way from being household names. When production at its Nebraska plant reaches full capacity, NatureWorks will still be making only 350 million pounds of its bio-plastic a year; Metabolix is able to produce about a third that much. Even if you count the various other biopolymers in existence and under development, the total barely registers in a world awash in fossil-fuel-based plastics.

Still, bioplastics are generating so much buzz because they're seen as holding the answer to many of our plastic woes — the rehabilitated partner who can transform this troubled relationship we're in. But before we commit to yet another family of polymers, it's worth asking the question Tim Greiner posed: Just what problems do bioplastics solve?

When I asked Michigan State polymer chemist Ramani Narayan, he had a single answer: the looming threat of climate change. Because biopolymers are made from "renewable sources of carbon" — what the rest of us call plants — they can reduce the amount of global-warming carbon dioxide we're sending up into the atmosphere. The CO_2 that's released at the bioplastic's end of life can be recaptured by the new plants that sprout in the next season. Bioplastics move us back inside the protective loop of the natural carbon cycle: the neatly balanced output and uptake of CO_2 that has sustained life on earth for eons. Even a single-use gift card that's immediately tossed in a landfill stays within that natural cycle if the card is made of a biopolymer, said Narayan. Petro-plastics, on the other hand, exist outside that loop, which is why their CO_2 emissions constitute a climate threat.

The benefit Narayan describes is quantified in the complex accounting of carbon calculations. Fossil-fuel-based plastics generate anywhere from two to nine kilograms of carbon dioxide for every kilogram of polymer produced. Plant-based plastics generate far less CO_2, even when you factor in all the oil used to fertilize, grow, and harvest the crops. PLA produces just 1.3 kilograms of carbon dioxide for every kilo of polymer produced. Mirel's carbon profile is a little higher because it takes more energy to make, but Narayan says it still beats conventional plastics.

Narayan has spent decades developing corn plastics. But he's not wedded to corn as a feedstock — any type of plant-based raw ingredient would have the same beneficial effect, he said. Indeed, agricultural crops — especially genetically modified crops — are probably not the best source of feedstocks. Critics point out the perversity of growing a food crop to make plastic in a world full of people who are hungry, and there's the additional drawback of the large amounts of land and water and oil-based fertilizers needed to raise it. A more sustainable and economical source of raw materials is waste — the vast quantities produced by us and the rest of the natural world. After all, conventional plastics are derived from the waste produced in processing fossil fuels; the ingenious use of that waste was what gave plastics an economic leg up in the first place.

Bioplastics producers are already exploring the possibilities posed by downed forest trees and the leftovers from paper and pulp production — surprisingly large sources of cellulose — as well as yard clippings and the remnants of harvested crops, such as corn stover and sugar-cane bagasse. By one estimate, such sources add up to 350 million metric tons a year, enough to substantially supplement fossil-fuel feedstocks. But there's also the waste we produce every day — our garbage, and even our sewage. Scientists around the world are dumpster-diving for waste materials that can be used as plastic feedstocks, exploring the possibilities of chicken feathers, orange peels, potato peels, carbon dioxide; even the methane emitted from landfills that is now sometimes recovered for energy. Stanford University chemist Craig Criddle is working with methane-eating microbes, relatives of the ones employed by Metabolix. He's found that after gorging on methane, they can produce prodigious quantities of a polymer similar to PHA and that the polymer can biodegrade back to methane. Though the technology is still in early stages, it promises a neatly closed production loop. So too does the work of Cornell University chemist Geoff Coates, who has developed a way to capture carbon dioxide emitted from the scrubbers on power plants and transform it into a biodegradable type of polypropylene carbonate plastic; it's now being commercially produced in small batches. None of these alone could replace the fossil fuels used in plastics, but there is no reason to believe we need just a single replacement. Part of petroleum's magic is that it has been able to do so many things so well. A more sustainable approach to plastics production (not to mention energy production) will almost certainly require developing multiple resources in the context of what is locally available and feasible. One measure of success will be how well bioplastic products, such as biopolymer credit cards, reduce an individual's carbon footprint.

But that's not the only plastic problem that biopolymers address. They also promise a safer chemical profile — certainly safer than the PVC typically used in credit cards. No hazardous compounds are required to knit a PLA or Mirel daisy chain. As Rossi noted, "I'd much rather live next to NatureWorks' plant in Nebraska than a petroleum-refining facility." And as producers of green plastics, NatureWorks and Metabolix have a vested interest in protecting how their resins are used and processed by manufacturers farther down the supply chain. "This sounds self-serving, but we are being held to a higher standard," said Davies. "From the get-go, we have to follow this extended producer responsibility model where we can't put products into the market without knowing where they go and what's happening to them."

Both NatureWorks and Metabolix claim to be committed to avoiding the addition of harmful chemicals by downstream processors and to being practitioners of the evolving science known as green chemistry. (Green chemistry goals include making synthetic chemicals with as few toxic substances and processes as possible, generating the minimum amount of waste, and producing compounds that won't persist in the environment.) NatureWorks, for instance, requires any manufacturer using its plastic to abide by a "prohibited substances list," which bars the use of various known persistent organic pollutants, endocrine disrupters, heavy metals, carcinogens, and other dangerous chemicals.

If you browse the shelves of bioplastic products, you'll notice that the most common problem they claim to address is plastic's stubborn durability. "Go ahead, throw it away! No composting required," boasts a maker of picnic forks, suggesting that once discarded, they will simply melt away. Fork gone; problem solved. But advertising, even the greenest, seldom tells the whole story.

I thought I knew what *biodegradable* meant, but in talking with experts, I came to realize it's a far more complicated process than my hazy notion of something just "breaking down." The term has a precise scientific meaning: *biodegradable* in this context means that the polymer molecules can be completely consumed by microorganisms that turn them back to carbon dioxide, methane, water, and other natural compounds. "The key word is *complete*," cautioned Narayan. It doesn't count as biodegradation if only a portion of the polymer can be digested.

That distinction is why Narayan has criticized my purportedly biodegradable Discover card. His studies show that despite the PVC-microbe bait, the micro-critters consume only about 13 percent of the card; after that, the process plateaus. It's also at the heart of a controversy over a rash of plastic bags that are marketed as "oxo-biodegradable." They're made of conventional plastics blended with an additive that causes them to break up when exposed to the sun. The bags do quickly crumble, but there's little evidence that the resulting plastic bits are ever fully consumed by microbes. Instead, critics contend, they may simply litter the earth with yet more tiny flakes of plastic.

Another complication affecting a product's biodegradability is that the process unfolds in different ways, depending on the material, the setting, and the microbes in residence. A felled tree is eminently biodegradable. In a steamy rainforest teeming with fungi and microbes, it could be gobbled up in a matter of months. Yet if it topples in the hot, dry desert where there are few microorganisms around, it will petrify long before it can be consumed. And if it sinks to the anaerobic bottom of a river, it will be preserved for centuries because the microbes that digest wood need oxygen to do their work. Plastics are intrinsically more difficult to break down than wood, but their capacity to biodegrade is a function of a polymer's chemical structure, not its starting ingredients. There are fossil-fuel-based plastics that will biodegrade (often used to make compostable bags and film), and there are plant-based plastics that won't.

In principle, both PLA and Mirel are biodegradable. In practice, it occurs more easily with Mirel. I could take a used Mirel gift card and toss it into my backyard compost bin, where microbes would digest it, creating lovely rich dark humus, over the course of a few months. The same would happen, though at a slower rate, if I lost it in the park, or even if I dropped it in the ocean. Mirel is just about the only plastic available today — petro- or plant-based — that will break down in a marine environment. So while you wouldn't want to build a dock with it, it could be a great material for plastic packaging, especially of foods and goods designed for shipboard use. Indeed, the U.S. Navy is exploring the use of Mirel utensils, plates, and cups.

PLA is trickier. It will biodegrade, but only under optimal composting conditions, which are challenging to achieve on one's own. Given the so-so state of my backyard compost bin, I suspect that if I deposited the PLA iTunes gift card there, it would remain intact for a good long while. Really mobilizing the microbes that can pry apart PLA's long polymer chains requires a balance of oxygen, moisture, aeration, and steady temperatures between 120 and 140 degrees — in short, the sort of conditions most readily found in an industrial composting facility. Unfortunately, there are only about two hundred to three hundred facilities in the country that process consumer food waste, and far fewer communities that actually collect residential food scraps for composting. Most of those are located in California and Washington.

As with any new technology, it takes time for a supporting infrastructure to develop. NatureWorks hopes that PLA products can eventually be chemically recycled, through a chemical process that breaks them back down to the starting ingredient, lactic acid. But as of 2010, there's only one facility in the world capable of doing that. At the moment, the plastic is creating a mini-crisis in the recycling world, where all is geared to conventional plastics. PLA is increasingly used for food packaging, but many consumers don't realize a PLA bottle can't go into the recycling bin. "We're freaking out about these," said one executive at San Francisco's Recology as he showed me a plastic water bottle made of PLA. The bottle looked exactly like one made of PET, yet it could contaminate a batch of PET being recycled. While some cup makers have started using green or brown logos and labels to indicate the cups are made of PLA, as of yet there's no standard system for differentiating biopolymers.

The allure of biodegradability is understandable. (Though it's ironic to see it assume the kind of marketing cachet for plastics that durability once held. I can't imagine any plastics maker today using this ad that ran in the 1980s: "Plastic is forever . . . and a lot cheaper than diamonds.") Still, the ability to biodegrade is neither a panacea for pollution nor the end-of-life solution to all things plastic.

Consider all the products, like that Discover card, that claim to break down in a landfill. It's a myth and a misplaced hope, said Steve Mojo; he's the director of the Biodegradable Products Institute, a trade group that polices the biopolymers world, certifying products that pass international standards of compostability and biodegradability. Ideally, nothing should biodegrade in a landfill, he explained. Landfills are engineered to deter that process as much as possible because it generates greenhouse gases. Yucky as it may be to think that our garbage will outlast us as well as our great-great-grandchildren, that's actually preferable to having it break down and give off methane, the most potent climate-change gas. Listening to Mojo describe how landfills work, I thought about the many biodegradable bags that are sold for collecting dog poop and that most people simply throw into the trash. These well-intentioned folks may be hoping that by their using biodegradable bags rather than regular plastic sacks, their pooches' poop will be more likely to decompose. But as with anything deposited in a landfill, "it's going to be preserved," said Mojo. "So when [future] generations go out and excavate the landfill, they will know we had a lot of dogs."

Where biodegradability makes sense is in products that are associated with food or organic waste (the sort that, unlike dog poop, can be safely composted), such as disposable plates and cups and cutlery, snack packages, and fast-food containers. All are single-use items that aren't often recycled today, especially the ones made of film. (Biodegradability would also be useful for the millions of pounds of agricultural film used by farmers every growing season to block weeds from sprouting among crops and that no one has found a way to economically recycle.) Making these kinds of products out of biodegradable bioplastics not only provides a solution for disposing of the package, it helps encourage the composting of food waste — which is a far bigger part of the garbage stream than plastics. Americans throw away more than thirty million tons of food waste each year, and most winds up in landfills. Zero-waste advocates see compostable plastic packaging as a two-for-one solution.

But is biodegradability the answer to the waste problems posed by quasi-cash plastic cards? Maybe. But what about redesigning them so that it's easier to load on new credit, allowing a card to be reused? That way, fewer new cards would have to be made. As for credit cards, why not reduce the frequency with which new cards are issued for existing accounts? Or expand on the few paltry card-to-card recycling programs that currently exist? Or make the cards out of a less toxic plastic than PVC so they can be more easily recycled? That's the route some European banks have gone and the one chosen by HSBC when it wanted to issue a more earth-friendly credit card for its Hong Kong market. Its green card, unveiled in 2008, is made from the most recycled plastic, PET. And it's backed by even more tangible ecobenefits: digital billing, which cuts down on paper waste, and the bank's pledge that a portion of all spending will be donated to local environmental projects.

Manufacturers have long chosen the plastics for their products on the basis of price and functionality. But creating a more sustainable relationship with plastics will require a new dexterity on our part. It will require us to think about the entire life cycle of the products we create and use. A green plastic that's suitable for one application may not be suitable for another when all environmental factors are taken into account. Biodegradation may not always be the best answer.

Consider the recent report in the *New York Times* that some designers of furniture and other housewares are taking pains to make sure their products are biodegradable. At one level, that's a laudable application of cradle-to-cradle thinking. Montauk Sofa, for instance, designed a line of couches in which all the components were made of organic, nontoxic materials that could biodegrade. As the chief executive of the company told the *Times,* "At first the whole idea was to have as little impact on the environment as possible. And then I started to think, wouldn't it be great to have no impact? Then it was, hey, what if the sofa just disappears when you're done with it?"

Leaving aside the question of whether that goal is even feasible, what does it say about our culture? Is a biodegradable couch a sign of a more sustainable mentality? Or is it just a greened-up version of the same old shop-and-toss habits? Traditionally, durability and longevity have bestowed additional value — a great-grandparent's walnut dresser isn't merely a place to store clothes; with time it becomes an heirloom, a connection to a past that has been conserved. Buying a two-thousand-dollar sofa designed for guilt-free disposal bears an uncomfortable resemblance to buying a ninety-nine-cent lighter also designed to be tossed. Wouldn't the lowest-impact sofa be one designed for and purchased with the expectation that it would be safely in use for decades?

Technology has come to define modern life, and we love the idea of gee-whiz technological fixes, even for the problems technology itself has created. Outrage at the Gulf oil spill is blunted by a fascination with high-tech blowout preventers and other technological marvels that promise to rescue us from our own complex creations. But the greening of Plasticville will require more than just techno- logical fixes. It also requires us to address the careless, and sometimes ravenous, habits of consumption that were enabled by the arrival of plastic and plastic money—a condition for which there is surely no better symbol than the maxed-out credit card. It means grappling with what historian Jeffrey Meikle called our "inflationary culture," one in which we invest ever more of our psychological well-being in acquiring things while also considering them of such low value "as to encourage their displacement, their disposal, their quick and total consumption."

What would it be like to turn your back on that culture—or at least the part of it involving plastic? I suppose I could have traveled to Lancaster, Pennsylvania, and spent time with an Amish family to find out. But instead I just picked up the phone and called Beth Terry, a fortysomething part-time accountant in Oakland, California, who in 2007 decided to start purging plastic from her life and is writing about her experiences on a blog she calls My Plastic-Free Life.

As Terry tells the story, she was home recuperating from a hyster- ectomy when she heard a radio report about Colin Beavan, a.k.a. No Impact Man, a New York resident who had pledged to live as lightly as helium for a year. Terry was moved by his story and decided to check out his blog. There she grabbed hold of an electronic chain that took her first to the (now-defunct) blog of Envirowoman, a Canadian woman who spent a year eliminating plastic from her life, and then to accounts of the plastic vortex, and then to the picture that she said changed her life: a photograph of a Laysan albatross carcass stuffed with plastic trash. The image tattooed itself onto her brain, forever altering her perspective on the world. "That bird was full of things that I use: it was bottle caps and toothbrushes and all the little pieces of plastic," she said. Looking at the photo, she was struck by how little control she had over things once they left her hands. Maybe, she said in hindsight, it was recovering from the hysterectomy, realizing she would never have children and being open to the idea of taking care of something else, like . . . the planet. Whatever the reason, she felt an urgent need to convert her horror into action.

She told me this story over lunch at an Oakland restaurant where we had arranged to meet. I had a feeling it was her when I saw the sensibly dressed woman with dark curls and wireless glasses push through the front door holding a cloth bag with the slogan Canvas Because Plastics Is So Last Year. The bag contained some of the accessories she carries with her to minimize her plastic intake, including cloth bags for the grains and produce she buys in bulk, as well as her kit for eating out: a wooden fork, spoon, and knife, in case she's presented with plastic cutlery; a pair of glass straws; and a cloth napkin. That day she was also toting a stainless steel pot, which she brought out when we later went to the butcher across the street to buy ground turkey for her cat (she is a vegetarian). In order to avoid the plastic film or plastic-coated paper used to wrap meat, she asked the butcher to put the ground turkey into the pot. I noticed she paid for it with a credit card. She says she doesn't have a problem using credit cards — the plastic lasts a long time — but she does worry a bit about the receipts because of the waste of paper and the fact that they are coated with bisphenol A. (Yet another of the ubiquitous chemical's uses: it bonds with the invisible ink used in carbonless copy paper to make an image appear when pressure, such as when one writes one's signature, is applied.)

As if I hadn't guessed it already, Terry explained she's not the sort of person who does things in half measures. When she took up running, she had to run a marathon; when she began knitting, she made scarves and hats for everyone she knew. So her goal of reducing plastic quickly went far beyond prosaic measures like using reusable bags and travel coffee cups. She began tracking the tiniest scraps of plastic that crossed her threshold — pieces of tape on packages received, the plastic windows in envelopes, the bits of film wrapped around the ends of organic bananas (a measure to prevent mold). She goes out of her way to rid herself of unwanted plastic: she's sent Tyvek mailers back to DuPont for recycling, returned the unneeded CDs that automatically were sent to her when she updated her version of TurboTax, and biked across town (she doesn't own a car) to take back Styrofoam peanuts to the shipper who had delivered a package from her dad. In all of 2009, she accumulated only 3.7 pounds of plastic—just 4 percent of the American average, she proudly noted on her blog. She cheerfully admits she's extreme but sees herself blazing a path that others can follow as far as they want to.

It's surprising how many people are game to try (though not her husband; he supports her efforts but hasn't joined her plastic-free crusade). Dozens of her readers have taken her up on her challenge to collect their plastic trash for a week or longer and then send in photos. In fact, the blogosphere is filled with plastic purgers and zero-waste zealots determined to reduce their footprints to the slightest tiptoe. They share recipes for homemade condiments and deodorant, fret over the frustrations of trying to find synthetic-free running clothes and sunscreen in nonplastic bottles, and swap tips for recycling unwanted plastic things such as gift cards. "Use them to scrape dried soy candle wax from tabletops, fabric, flat candleholders," one of Terry's readers suggested. "Use them to crease folds in papercrafting . . . [C]ut them into squares, glue them onto cork, and make coaster mosaics." They confess their consumption sins online—"Out of laziness, I broke down and bought tortillas in plastic," one reader wrote Terry.

Even among this hard-core crowd, there are levels of extreme. A fellow green blogger accused Terry of "hair shirt environmentalism" for using baking soda and vinegar to wash her hair. This, noted Terry, from a woman who advocated using cloth wipes in place of toilet paper, "which I think is really extreme." But to Terry, it didn't feel like any great sacrifice to give up bottled shampoo in favor of baking soda and vinegar. It's cheaper, which appeals to her frugal nature. Besides, she added, "I'm not very girly and never have been." (Envirowoman, one of the first blogging plastiphobes, complained regularly about the difficulty of finding plastic-free cosmetics.)

"Is there anything you've done that *does* feel like hair-shirt environmentalism?" I asked.

"I miss cheese." She laughed wistfully. The sharp cheddar she likes almost invariably comes wrapped in plastic. Eventually she managed to find a cheese — not cheddar, alas — wrapped in natural beeswax. But she had to buy the entire fifteen-pound wheel. Occasionally she tries to give herself a break from herself. "I went to Trader Joe's the other day just to get something quick for lunch. I used to be able to eat at Trader Joe's all the time. I wanted to get a salad." She was fully prepared to confess the transgression in her next blog entry. But then that image of the plastic-stuffed albatross flitted across her mind. "I just couldn't do it. I looked at all the plastic and just walked out."

Over the course of her deplasticization, she's had to abandon purchases ever more frequently. As Terry recalled, at first she simply wanted to replace plastic stuff with things made of glass or wood or paper or other natural materials. She bought sauces in glass jars, scoured the grocery stores for frozen dinners that came in nonplastic trays, tried soy milk powder to make soy milk (pronouncing it "feh!"), and gave up disposable razors in favor of an old-fashioned safety razor she found at a local antique store.

"I thought I could find an alternative for everything in my house," she said. But over time, she found that "there were fewer and fewer things I could buy." When her hair dryer broke, she had to go without or figure out a way to repair it, which she did. Instead of buying almond milk and yogurt and cough syrup, she taught herself how to make them. Rather than purchase new tools, she borrowed them from friends or a local tool-lending program.

"Giving up plastic," Terry said she realized, "meant I was kind of forced to consume less." She may not have environmental quibbles with her plastic credit cards, but the fact is, a life without plastic means she has fewer and fewer occasions to use them.

Plastic is so deeply embedded in our consumer culture it is almost synonymous with it. Look at the bright, shiny hygienic surface of Plasticville and you'll see a wealth of products that make life easier, more convenient. But start scratching that surface and you'll begin to see that minor, even trivial, conveniences can have profound consequences — whether that's reflected in disposables that will outlive us, chemicals that can undermine the health and fertility of future generations, or albatrosses choking on things we've discarded because they can't be reused or recycled.

Does this mean we must follow Terry down that road out of Plasticville? Must we choose between our plastic and our planet? If those were the only options on offer, I'm not sure I could trust myself or my fellow citizens to make a good decision. Fortunately, building a sustainable future doesn't require such a stark and dramatic choice. In fact, an overly simplistic pursuit of perfection can get in the way of a mostly green good.

Consider local dairies trying valiantly to improve on the way milk is produced and sold. One in my area sells organic milk in returnable glass bottles. But the cap is still plastic, and for Terry that's a deal breaker. It's a question of priorities, she said. "You have to prioritize what's important in your life. I don't need to drink milk." That may be a reasonable choice for Terry, but if enough people followed her example, that organic dairy with its returnable glass bottles would go out of business. If we want to bring about a greener world, personal virtue must take into account the larger political and social contexts of individual actions. Still, Terry's uncompromising example provides a reminder of the tradeoffs we casually make every day, as I realized when I finally decided to take up her plastics challenge and track my plastics consumption for a week.

I'd been putting it off. I'm not sure why, except the whole idea made me feel vaguely uncomfortable. I knew there was no way I was going to scale back plastics to the degree Terry had. I have three kids, full-time work, and a far less obsessive temperament; I've never felt compelled to run a marathon. I wasn't convinced that collecting my plastic trash for a week would tell me anything I didn't already know. Or, if I'm honest, anything I wanted to know.

To my surprise, it turned out to be a very useful exercise, like my earlier experiment in writing down everything I touched that was plastic. It reminded me once again of plastic's ubiquity and how easy it is to stop noticing that fact. Knowing that I would have to keep and consider every plastic item I used transformed each use — even the most trivial — into a conscious decision. At the gym, I could get myself a drink of water from one of the plastic cups by the cooler — and add that cup to my collection. Or I could walk downstairs and sip from the water fountain.

Looking at the pile of trash I accumulated in a week — 123 items, which was probably more than Terry generated in a year — a few things became clear. One was how often my purchases are made on the basis of convenience. Do I really need to buy zucchini from Trader Joe's, where it comes nestled on a plastic tray, covered in plastic wrap, with little plastic stickers adorning each individual squash? Sometimes. But most weeks I can make the time to stop by the farmers' market or the neighborhood produce stand, where all the fruits and vegetables come unencumbered by synthetic skin.

I was embarrassed to realize how many of the packages I'd collected that week contained food that had gone bad because we hadn't finished it. There were five bread bags, each of which held a few moldy slices — the dreaded heels of the loaf that my kids refused to eat. Those bags were evidence that I was doing far too much of my grocery shopping on autopilot, without thinking carefully about what we really need. But it also reminded me of something Robert Lilienfeld, the coauthor of *Use Less Stuff,* told me when I spoke with him about the debate over plastic shopping bags. He pointed out that for all the environmental troubles single-use shopping bags cause, the much greater impacts are in what they contain. Reducing the human footprint means addressing fundamentally unsustainable habits of food consumption, such as expecting strawberries in the depths of winter or buying varieties of seafood that are being fished to the brink of extinction.

Beth Terry's challenge pinched awake my sense of mindfulness about my grocery shopping, reminding me to ask myself as I wheeled my cart through the store: Is this something we really need? I'm going to answer that question "yes" more often than Terry. But it's never a bad question to ask oneself, especially in a consumer culture that encourages people to swipe their credit cards regularly but not necessarily thoughtfully.

Those credit cards provide a powerful way to help shape the

choices consumers are offered. We can use them to vote for healthier, safer products and to support the development of plastics that are genuinely green. We can also vote by keeping them firmly tucked inside our wallets and rejecting overpackaged goods and products that can't be reused or recycled. The power of the purse has helped make sustainability a viable niche in the market, fueling sales in durable water bottles, travel mugs, and the like. It's why Walmart now sells organic produce and why Clorox introduced a toxin-free line of cleaning products and why the makers of baby bottles and sports water bottles voluntarily switched to bisphenol A–free alternatives. We can move markets, as Terry demonstrated in 2008 when she organized a successful campaign to get Clorox to recycle the carbon cartridges used in its Brita water filters — something the European maker of Brita had begun doing years before, thanks to the requirements of extended-producer-responsibility laws.

But individual actions alone are unlikely to bring about change on the scale that is now required — whether the task is stopping the plasticization of our oceans, protecting our children from endocrine disrupters, or curbing the carbon emissions that fuel global warming. The forces that shaped our marriage with plastics — a powerful petrochemical industry, a culture of acquisition, an erosion of community-mindedness in the suburban diaspora — evolved in a political culture that assumed a world without biological limits. That genie can't be put back in the bottle, but we can remold our political culture to make the genie a better citizen.

Government at all levels — from city councils to Congress — has a role to play in reinventing our communities as places where it is easy, convenient, and cost-effective for people to use less, reuse more, recycle, and compost; where businesses that serve those ends can thrive; where all producers take cradle-to-cradle responsibility for the things they create; and where the ocean is valued for the vast resource it is rather than being the final dumping ground of our plastic folly.

It's a huge project, remaking our relationship with this family of materials.

We've produced nearly as much plastic in the last ten years as we have in all previous decades put together. We've become used to our polymer partners, for better and worse. Today's college graduates may not want a career in "Plastics!" any more than Dustin Hoffman did, but their lives are going to be defined by the presence of plastics to a greater degree than the lives of any previous generation. Plastic production is accelerating, plastic goods are spilling out across the landscape, a culture of use-and-dispose is being exported to a developing world whose consumption of plastic could, by some estimates, catch up to U.S. and European levels in the next forty years. Our annual global plastics production, if present trends hold, could reach nearly two trillion pounds by 2050. If it feels like we're choking on plastic now, what will it feel like then, when we're consuming nearly four times as much?

We have come a long way from the early promise of plastics, a substance we hoped could free us from the limits of the natural world, democratize wealth, inspire the arts, enable us to make of ourselves virtually anything we wanted to be. But for all the wrong turns we've taken, plastic still holds out that same promise. Especially in a world of seven billion souls — and counting — we need plastics more than ever. We have to remind ourselves that our power to create a sublime world resides not in the materials we deploy but in our gift for imagination, our capacity to create community, our ability to recognize danger and to seek a better way.

Just as individual action is no substitute for the exercise of our collective political will, neither can we simply legislate our way to that sustainable, enriching future we know is possible. Remaking Plasticville into a place where our children and their children and their children can safely live will require us to confront assumptions about ourselves and what we need for fulfilling lives and satisfied minds. We don't need to reject material things but to rediscover that their value may reside less in the quantity of things we own and — as with Della's comb — more in the way our material possessions connect us to one another and to the planet that is the true source of all our wealth.

A Bridge

The bridge is unremarkable-looking—just a short, plain span connecting one dirt road to another deep in the heart of the New Jersey Pine Barrens. Pitch pines, scrub oaks, and black gum trees line the road leading up to it. Blueberry and leatherleaf bushes cover the riverbanks on both sides. It's one of dozens of bridges that crisscross the tea-colored waters of the Mullica River as it winds its way through a woods called Wharton State Forest. Unlike the other bridges, however, this one is made entirely of plastic.

You wouldn't necessarily know that unless you stopped to give it a good once-over. All the same, a spokeswoman for the state forest told me, "It doesn't look out of place." In fact, because it's made entirely from recycled plastics, she said, "it promotes our focus of being green."

Nearly one million used milk jugs and a lot of old car bumpers were smooshed and melted and remolded to make the plastic I-beams, pilings, and planks that were used to construct the fifty-six-foot-long bridge. Rutgers University polymer scientist Thomas Nosker invented the technology to turn plastic throwaways into durable building materials. He then licensed it to a New Jersey company, Axion

International, that's taking it commercial. Axion says the plastic it produces from recyclables can be molded to make bridges, railroad ties, decks, pilings, bulkheads, and levees and will stand up to time and the elements far better than wood or concrete or steel. In just two years the company has created worthy new lives for more than two million pounds of plastics that might otherwise have wound up in a landfill. For Axion founder Jim Kerstein, these kinds of products are karmic payment for a career he spent producing and selling hangers made of virgin plastics that he knew would almost invariably be thrown away. "All the negatives about plastic — that it lasts long and doesn't degrade, are being turned into positives," he said. "You're taking a material that doesn't degrade and putting it to use where we want it to last forever."

The Wharton State Forest bridge, constructed in 2002, was one of the company's first. Intrigued by the technology, the U.S. Army hired the company to put up a pair of bridges across small creeks at Fort Bragg. Axion promised that these bridges would support not only trucks but also M1 Abrams tanks, which weigh in at seventy tons each. Army engineers were so dubious that a plastic structure would support a tank that they brought along a crane to the tank's test run across the bridge, convinced they'd need to haul the tank from the creek. The monster tank rumbled across the twenty-foot span, and the bridge scarcely flexed. "Others build strong bridges, but this bridge was built Army-strong," an admiring representative from the U.S. Army's office of acquisition declared at the bridge's dedication in 2009.

The military, he noted, spends $22.5 billion a year on replacing structures lost to corrosion. This bridge cost less to build than those made with other materials and will be corrosion-resistant and practically maintenance-free. After the Fort Bragg bridges were finished, the army ordered two more for its base in Fort Eustis, Virginia. These bridges are slated to carry railroad locomotives weighing 120 tons.

The deteriorating wooden bridge that Axion replaced at Wharton State Forest was at least fifty years old; it had been there when the Wharton family granted the land to the state, in 1954. The plastic bridge is likely to last much, much longer. Barring an act of war or a natural calamity, it will be there long after the nearby oaks and pines have toppled and new trees have taken their place, ready to be crossed by distant generations awaiting their turn on the planet. So often plastic's obdurate persistence is an insult and injury to the natural world. But this modest crossing in the middle of the forest seems just the right application of a material that doesn't die. Maybe it's not right for a span like the Tappan Zee or the Golden Gate Bridge. But, said Kerstein, most of the six hundred thousand or so bridges in the United States are small spans — shorter than seventy feet — in which traditional materials could easily be replaced by recycled plastic.

The oldest bridge in the world is the Arkadiko Bridge in southern Greece. Three thousand years ago, masons fit together rough limestone boulders to form a simple arch about twelve feet high and sixty feet long over what once was a river and now is a dry, grassy gully. Looking at the ancient structure, you can almost hear the clatter of horse-drawn chariots crossing in their travels between Mycenaean cities. The Arkadiko dates from the late Bronze Age, which came to an end in a catastrophic collapse that has been variously attributed to volcanic eruptions, earthquakes, raids by other cultures, and climate change.

Today, for better and for worse, we are firmly in the plastics age and facing frightening intimations of ecological collapse. We have at hand the materials to help avert it, tools with which to create a legacy of sustainability. Will archaeologists millennia from now scrape down to the stratum of our time and find it simply stuffed with immortal throwaways like bottle caps, bags, wrappers, straws, and lighters — evidence of a civilization that choked itself to death on trash? Or will they come upon bridges like the one in Wharton State Forest, bridges that, despite their lack of beauty, have an important story to tell: that we were a people with the ingenuity to make wondrous materials and the wisdom to use them well.

Cast of Characters

The plastics we're most likely to encounter — in approximate order of commonness.

Polyethylene: If there were a popularity contest for polymers, polyethylene would win hands-down. More than a third of all plastics produced and sold worldwide belong to this plastics family. They're tough, flexible, moisture-proof, and exceptionally easy to process, which make them favorites for packaging. Members of the clan include:

- **Low-density polyethylene (LDPE):** used to make bags (for newspapers, dry cleaning, frozen foods, and so forth), shrink-wrap, squeezable bottles, coatings on milk cartons, and hot and cold beverage cups
- **Linear-low-density polyethylene (LLDPE):** a stretchier version of polyethylene, used in bags, shrink-wrap, lids, pouches, toys, and flexible tubing
- **High-density polyethylene (HDPE):** a hardier variety of polyethylene, used for those ubiquitous plastic grocery bags as well as

Tyvek home insulation. In a stiffer incarnation, it's used for the bottles containing milk, juice, detergent, and household cleaners, and for the bags in cereal boxes.

Polypropylene: Though related to polyethylene, this plastic can handle higher temperatures and rougher treatment, giving it a different niche in packaging. It can take the stress and twisting required of bottle caps and hinge-top lids. Its high melting point makes it useful for the reusable containers sold for leftovers; for containers that are filled with hot contents, such as freshly boiled syrup; for incubating yogurt; and for takeout foods. Cars are chock-a-block with polypropylene, inside and out—from the bumpers to the carpeting to the substrate beneath interior fixtures. In textile form, polypropylene allows moisture to escape while staying dry, making it useful in disposable diapers, thermal vests, and even astronauts' space suits.

Polyvinyl chloride (vinyl): One of the most versatile (and controversial) of all plastics. PVC can take on a staggering variety of personalities—rigid, filmy, flexible, leathery—thanks to the ease with which it can be blended with other chemicals. Vinyl surrounds us at every turn. We use it to side houses; to cover walls, floors, and ceilings; to insulate electrical wires; as "pleather" clothing and Naugahyde upholstery; as the housing and jointing for pipes; and as the pliable plastic in medical devices.

Polystyrene: Most recognizable when puffed up with air into that synthetic meringue known technically as expanded polystyrene and popularly by the trademark Styrofoam. In that guise, it's an excellent insulator—of homes, hot coffee, an order of chow mein, a fragile shipment, our heads (while we're biking). But it can also take the form of a strong, hard plastic, deployed for CD jewel cases, videocassette cartridges, disposable razors, and cutlery. A high-impact version is used for coat hangers, smoke-detector housings, license-plate frames, aspirin bottles, test tubes, petri dishes, and model-assembly kits.

Polyurethane: Introduced in 1954, polyurethanes are a big family of plastics that come in foamed versions that are variously soft and flexible (think of the cushioning in furniture, cars, and running shoes); tough and rigid (the insulating lining for buildings and refrigerators); and somewhere in between (the padding on dashboards). Polyurethane's supportive abilities take on a whole other meaning when it is spun into a fiber to make Spandex or Lycra or extruded into a thin film for latex-free condoms.

Polyethylene terephthalate (PET): The most prominent member of the polyester family, PET debuted as a wrinkle-proof fiber introduced after World War II, and textiles are still the end use of most polyester. PET soon gained other uses: in photographic and x-ray films, audio- and videotapes. But its biggest claim to fame has come from the benefits it brings to packaging: a glasslike clarity and an unparalleled ability to provide an airtight container, keeping food-spoiling oxygen out and fizz-producing carbon dioxide in. Now almost every kind of beverage comes in PET — possibly even your next bottle of wine.

Acrylonitrile butadiene styrene (ABS): ABS was created in the 1940s by scientists trying to make synthetic rubber. The resulting co-polymer of three starting ingredients is a hard, glossy, shock-absorbing, distinctly unrubberlike material that's the stuff of Legos; musical instruments such as recorders and plastic clarinets; golf-club heads; casings for phones and kitchen appliances; automotive body parts; and other light, rigid molded products.

Phenolics: This family of polymers is descended from the first wholly synthetic plastic, Bakelite. Unlike other common plastics, phenolics cannot be melted and remolded. Strong, hard, capable of insulating against electricity, these plastics are used in electrical fixtures and switch gears, as well as in Formica and cutlery handles. Bakelite, the best-known phenolic and once a plastic that touched every corner of life, has now been mainly sidelined to the world of games, where it's a favorite for making chess pieces, checkers, dominoes, and mahjong tiles.

Nylon: The name is a DuPont trademark that covers a diverse class of plastics. The same qualities that revolutionized women's stockings — strength, durability, and elasticity — make nylon desirable for a host of other stuff. Born from the search for artificial silk, nylon fibers are used for fabrics, bridal veils, musical strings, carpets, Velcro, and rope. In solid form, other types of nylon are used for machine screws, gears, boat propellers, combs, skateboard wheels, fuel lines and fuel tanks, and the bristles in toothbrushes and hairbrushes.

Polycarbonate: A family of engineering plastics, polycarbonate was developed to compete with die-cast metal. It can be one of the toughest of plastics, but it's also transparent, a combination that's made it a choice for use in gears, compact discs, DVDs, Blu-ray discs, eyeglass lenses, lab equipment, the housing for power tools, and, until recently, containers that you don't want to have shatter, such as baby bottles and sports water bottles. But concerns about the plastic's tendency to leach the chemical bisphenol A have all but eliminated those last uses from the marketplace.

Acrylic: Clear as glass, but immeasurably tougher, acrylic can stand up to harsh weather and stop bullets, which is a combination that made it ideal for protecting airborne gunners during World War II. These days it's gained service in the presidential motorcade, aboard the Popemobile, and in the teller windows at drive-through banks. But acrylic also reports for many less glamorous duties: airplane windows and submarine ports, outdoor signs and car taillights, residential and commercial aquariums, replacement lenses for cataract sufferers, and as a stand-in for glass shower doors.

Acknowledgments

When I proposed writing an all-around look at the world of plastics, I had no idea how enormous the project would be. Just as plastics reach far into many aspects of modern life, so did my research, taking me into dozens of fields about which I initially knew very little. So I got a lot of help assembling this literary daisy chain. Acknowledging every person I spoke with would take up a second volume. But in addition to those already mentioned in the book, I am indebted to the following for generously sharing their time and expertise:

Raymond Giguere of Skidmore College provided the basic chemistry lessons I slept through in high school. In Leominster, Massachusetts, I had the great good fortune to meet with Louis Charpentier, who began working in plastics in 1927 and who shared with me both his recollections of the industry's early years and his collection of buttons. Marianne Zephir, former curator for the National Plastics Museum in Leominster, which, sadly, closed in 2009, was also helpful to my early research. I was lucky to get insights about the early years of plastic from Julie Robinson, a historian of celluloid, and Don Featherstone, creator of the pink flamingo.

For basic tutelage in polymer technology, my thanks to Jeffrey Wooster and Bob Donald, of Dow Chemical Company; Dan Schmidt, University of Massachusetts at Lowell; and Matt Naitove, *Plastics Technology*. For a glimpse of the wild possibilities polymers offer, my appreciation to Material ConneXion and staffers Beatrice Ramnarine and Cynthia Tyler.

For sharing industry perspectives on various plastics-related issues: Chris Bryant, Keith Christman, Steve Hentges, Jennifer Killinger, Steve Russell, and Tim Shestek of the American Chemistry Council; Glenn Beall, Glenn Beall Consulting; Robert Bateman, Roplast Industries; Isaac Bazbaz, Superbag Corporation; Mike Biddle, MBA Polymers; John Burke, Food Service Packaging Institute; Bill Carteaux, Society of the Plastics Industry; Mark Daniels, Hilex Poly Company; David Durand, Townsend Solutions; Marc Greene, Axion International; David Heglas, Trex Inc.; Kevin Kelley, Emerald Packaging, Inc.; Tony Kingsbury, Dow Chemical; George Mackinrow, consultant and author of an unpublished account of the bag wars; Robert Malloy, chair of polymer engineering at the University of Massachusetts at Lowell; Ken Pawlak, author of a forthcoming book on plastics in medicine; Alicia Rockwell, Savemart Corp.; and C. A. Webb, Preserve, Inc. Rafael Auras, Diane Twede, and Susan Selke, of the Michigan State University School of Packaging, helped me better understand the science and technology of packaging. I gained a deeper appreciation for the serious business behind the fun of the toy industry with the help of Bill Hanlon, Learning Mates; Robert von Goeben, Green Toys; Dan Mangone, Discovering the World; Sally Edwards, University of Massachusetts at Lowell, Center for Sustainable Production; Tim Walsh, author of *Wham-O Super-Book: Celebrating Sixty Years Inside the Fun Factory*. Kelly Chapman and Stan Chudzik of Goody's were helpful sources of information about the business and technology of combs.

Art and design were new fields to me, happily made more comprehensible thanks to help from George Beylerian, Material ConneXion; Cristiano de Lorenzo, Christie's in London; and Alexander von Vegesack, Vitra Design Museum. Also thanks to Manfred Dieboldt, a former engineer with Vitra, for his recollections about the making of the Panton chair.

My visit to China would have been all for naught without the help and insights of Fu An, Guangdong Plastics Industry Association; Joe Wong, Innotoys; Victor Chan, Wild Planet; Tony Lau, Canfat Manufacturing; L. T. Lam, Forward Winsome Industries; Sarah Monks, chronicler of the Hong Kong toy industry; Jurvey Gong, ColuComan Chemical Technology; and the invaluable assistance of Steve Toloken of *Plastics News.*

For help in explicating the complex mechanisms of endocrine disrupters and the health concerns posed by the use of plastics in medicine, my gratitude to George Bittner, University of Texas at Austin; John Brock, Warren Wilson College; Antonia Calafat, Centers for Disease Control; Earl Gray, Environmental Protection Agency; Russ Hauser, Harvard School of Public Health; Patricia Hunt, Washington State University; Mark Ostler, Hospira, Inc.; David Rosner, Columbia University; Ted Schettler, Science and Environmental Health Network; Rebecca Sutton, Environmental Working Group; and Sarah Vogel, author of a forthcoming book on the politics of bisphenol A. Thanks also to the following health-care administrators and practitioners for conversations about the roles and risks of medical plastics: Valerie Briscoe, John Muir Medical Center; John Fiascone, Tufts University; Julianne Mazzawi and Irena Solodar, Brigham and Women's Hospital NICU; Gina Pugliese, Premier, Inc.; Laura Sutherland, Practice Green Health; and Susan Vickers, Catholic Healthcare West. And my appreciation to attorneys Billy Baggett, of Lake Charles, Louisiana, and Herschel Hobson of Beaumont, Texas, for describing their experiences in vinyl chloride litigation.

To get up to speed on the complicated science of oceanography and marine debris, I got assistance from Joel Baker, University of Washington; James Dufour, Peter Niiler, and Miriam Goldstein of Scripps Institution of Oceanography; Holly Bamford, National Oceanographic and Atmospheric Administration; David Barnes,

British Antarctic Survey; James Ingraham, formerly of NOAA; Kara Lavender Law, Sea Education Association; and Hideshige Takada, Tokyo University of Agriculture and Technology. A particular thanks to the committed folks associated with Project Kaisei, who spent many hours discussing their work with me: Mary Crowley, Michael Gonsior, Andrea Neal, George Orbelian, Dennis Rogers, and Douglas Woodring.

I am also indebted to a number of political activists and policy experts who explicated the intricacies of various plastics-related issues. Their ranks include Vince Cobb, ReusableBags.com; David Allaway, Oregon Department of Environmental Quality; Lindy Coe-Juell, City of Manhattan Beach; David de Rothschild, Plastiki Expedition; Bryan Early, Californians Against Waste; Marcus Erikson, Algalita Marine Research Foundation; Mark Gold, Heal the Bay; Miriam Gordon, Clean Water Action; Joe Greene, Chico State University; Richard Lilly, Seattle Public Utilities; Cheryl Lohrmann, founder of Leave No Plastic Behind; Pam Longobordi, plastic-beach-debris artist; Brady Montz, Sierra Club in Seattle; Heidi Sanborn, California Product Stewardship Council; Sharron Stewart, activist in Lake Jackson, Texas; Leslie Tamminen, Seventh Generation Advisors; Emily Utter, formerly of Chico Bags; Michael Wilson, University of California at Berkeley. Tim Kasser of Knox College offered useful insights on the psychology of consumerism. Mike Verespej of *Plastics News* was a tremendous source of help on the bag fights and other policy debates concerning plastics.

For help in explaining what happens to my plastic things once I'm done with them, I'm grateful to the following experts on recycling and waste disposal: Frank Ackerman, Tufts University; Lyle Clark, Stewardship Ontario; Susan Collins and her predecessor Betty McLaughlin at the Container Recycling Institute; Paul Davidson, WRAP; Edward Kosior, Nextek Ltd.; Charlie Lamar and Ed Dunn, Haight Ashbury Neighborhood Council Recycling Center; George Larson, solid-waste consultant; Ted Michaels, Energy Recovery Council; Patty Moore, Moore Associates; Clarissa Morawski, CM Consulting; Bruce Parker, National Solid Waste Association; Jerry

Powell, editor, *Resource Recycling and Plastics Recycling Update;* Robert Reed and Leno Bellomo, Recology; Dennis Sabourin, National Association for PET Container Resources; Alan Silverman, Eagle Consulting; Peter Slote, Oakland Solid Waste and Recycling; Kit Strange, Resource Recovery Forum; and Kathy Xuan, Parc Corporation.

My guides to the brave new world of postpetroleum plastics included Tilman Gerngross, Dartmouth College; Brian Igoe, Metabolix; Brenda Platt, Institute for Local Self-Reliance; and Frederic Scheer, Cereplast.

I'm also grateful to the sustainability gurus who helped me parse the many meanings of that overused word: John Delfausse, Estée Lauder; Ann Johnson of the Sustainable Packaging Coalition; Robert Lilienfeld, coauthor of *Use Less Stuff: Environmentalism for Who We Really Are;* Andrew Wilson, author of *Green to Gold;* Cathy Crumbley and Ken Geiser, of the Lowell Center for Sustainable Production, and huge thanks to the team of experts from the center, the Toxics Use Reduction Institute, and the members of UMass Lowell's polymer engineering department, who gathered together for a day to let me pick their brains: Pam Eliason, Greg Morose, Liz Harriman, and Ramaswamy Nagarajan.

The following folks were kind enough to read and provide feedback on various parts of this manuscript: Susan Collins; Cathy Crumbley; Peter Fiell; Robert Friedel; Robert Haley; Russ Hauser; Marianna Koval; Naomi Luban; Donald Rosato; Dan Schmidt; Seba Sheavly; Shanna Swan; Brenda Platt; Jerry Powell; Joel Tickner; and Nan Wiener. Any errors that remain are wholly my own.

Of course, all that expert input would hardly add up to a coherent whole without strong editorial guidance and boatloads of support. I was incredibly lucky to have both, starting with my fabulous agent Michelle Tessler and my extraordinary editor Amanda Cook, who is the kind of engaged, insightful, demanding reader that writers dream of but rarely encounter in publishing these days. Thanks also to ace copyeditor Tracy Roe for rescuing me from multiple embarrassments in style and grammar and to Lisa Glover for shepherding the book through production. My sister, Lisa Freinkel, helped develop the organizational schema of the book and absorbed far more plastic trivia than she ever wanted to know. My brother, Andrew Freinkel; my mother, Ruth Freinkel; and my three children, Eli, Isaac, and Moriah Wolfe all provided vital support. I'm lucky to have a book finder in the family, so am very grateful to my brother-in-law Ezra Tishman for sending obscure polymer texts my way. Also, many thanks to Leslie Landau and Jim Shankland for once again letting me use their beautiful guesthouse for a much-needed writing retreat. I can't imagine even trying the long slog of book writing without my wise and wonderful writing group, North 24th — Allison Bartlett, Leslie Crawford, Frances Dinkelspiel, Sharon Epel, Kathy Ellison, Katherine Nielen, Lisa Okuhn, Julia Flynn Siler, and Jill Storey. Above all: bottomless appreciation for my husband, Eric Wolfe, for his support, patience, enthusiasm, and amazing capacity to find the meaning of it all.

Notes

Introduction: Plasticville

page

1 *a place it called Plasticville:* Jeffrey Meikle, *American Plastic: A Cultural History* (New Brunswick, NJ: Rutgers University Press, 1997), 189. The following website contains information on the toys: http://www.tandem-associates.com/plasticville/plasticville.htm.

4 *a "petting zoo":* George Beylerian, quoted in John Leland, "The Guru of Goo (and Gels, Mesh, and Resin)," *New York Times,* March 14, 2002.

5 *"Nylon . . . the Gay Deceiver": House Beautiful* 89 (October 1947): 161.
 You might as well claim: Robert Kanigel, *Faux Real: Genuine Leather and Two Hundred Years of Inspired Fakes* (Washington, DC: Joseph Henry Press, 2007), 87.
 Chains crowded close: Herman F. Mark, *Giant Molecules* (New York: Time-Life Books, 1966), 64.

6 *"a fourth kingdom":* Quoted in Meikle, *American Plastic,* 114.
 You could also peg the dawn: J. Harry DuBois, *Plastics History, U.S.A.* (Boston: Cahners Books, 1972), 197.
 Styrofoam: The word is trademarked by Dow Chemical to refer to a plastic technically called expanded polystyrene. But I use it here to refer to any product made of that foamed polystyrene.
 "virtually nothing was made from plastic": Quoted in Meikle, *American Plastic,* 180.

7 *Oil refineries run 24-7:* Barry Commoner, foreword to Kenneth Geiser, *Materials Matter: Toward a Sustainable Materials Policy* (Cambridge, MA: MIT Press, 2001), x–xi.

"By its own internal logic": Barry Commoner, *The Poverty of Power: Energy and the Economic Crisis* (New York: Random House, 1976), quoted in Meikle, *American Plastic,* 265.

amount of plastic the world consumes: "FYI: Global Plastics Resin Production Over the Years," *Plastics News,* October 30, 2009. In 2008, total production was 540 billion pounds, down from 573 billion the previous year—a reflection of the recession.

In 1960, the average American: Dominick Rosato et al., *Markets for Plastics* (New York: Van Nostrand Reinhold, 1969), 3.

Today we're each consuming: The per capita figure is based on a U.S. population of three hundred million and annual plastics production of a hundred billion pounds, which is a little less than the amount produced in 2008 and slightly higher than the postrecession production in 2009. In 2009, sales were $327 billion, a significant drop from the $374 billion in 2007, reflecting the recession. Society for the Plastics Industry, "Fast Facts on Plastics and the Economy." Also, Robert Grace, "SPI Reports on US Plastics Market Recovery," *Plastics News,* October 28, 2010.

8 *Considering that lightning-quick ascension:* Author interview with Frederic Scheer, president of Cereplast, Inc., May 2010.

It's "wonderful how du Pont": Meikle, *American Plastic,* 135.

A few years later, people told pollsters: Geoffrey Nunberg, *Going Nucular: Language, Politics, and Culture in Confrontational Times* (New York: Public Affairs, 2004), 4–5; author e-mail correspondence with Nunberg, March 2010.

a new era of material freedom: Robert Friedel, *Pioneer Plastic: The Making and Selling of Celluloid* (Madison: University of Wisconsin Press, 1983), 28.

a "universal state": Edwin Slosson, *Creative Chemistry* (New York: Century, 1919), 132–35, quoted in Meikle, *American Plastic,* 70.

"I just want to say one word": The endurance of that line has been a source of dismay to the plastics industry, whose main trade journal "could not bring itself to refer to that 'tired old joke about plastics' until 1986." Meikle, *American Plastic,* 3.

9 *"a malign force"*: Quoted in Stephen Fenichell, *Plastic: The Making of a Synthetic Century* (New York: Harper Collins, 1996), 306.

Humans could disappear: The point was elegantly explored by Alan Weisman in his book *The World Without Us* (New York: Thomas Dunne Books, 2007).

10 *As historian Robert Friedel notes:* Robert Friedel, "Some Matters of Substance," in *History from Things: Essays on Material Culture,* Steven Lubar and W. David Kingery, eds. (Washington, DC: Smithsonian Institution Press, 1993), 49–50.

plastic in the first decade: Richard C. Thompson et al., "Plastics, the Environment, and Human Health: Current Consensus and Future Trends," *Philosophical Transactions of the Royal Society B* 364 (July 2009): 2166. The entire issue of the journal was devoted to plastics and the health and environmental problems associated with them. It's an excellent overview of current scientific understanding and concerns.

1. Improving on Nature

13 *"the necessities and luxuries of civilized life"*: Edward Chauncey Worden, *Nitrocellulose Industry,* vol. 2 (New York: D. Van Nostrand, 1911), 567, quoted in Meikle, *American Plastic,* 15.

14 *"Have you ever seen a polypropylene molecule?"*: Author interview with Bill Adams, owner of Adams Manufacturing Corporation, March 2008.

15 *at least one million pounds of ivory:* "The Supply of Ivory," *New York Times,* July 7, 1867. Historian Robert Friedel argued that fears of the shortage were overblown — the only truly scarce commodity was the particular type of ivory required for billiard balls, which was taken from the center of select tusks.

In 1863, so the story goes: The story of Hyatt and celluloid's early days is told in Fenichell, *Plastic,* 38–45, and Meikle, *American Plastic,* 10–30. Some, especially in Britain, consider the true father of plastics to be Alexander Parkes, an English inventor who was the first to combine cellulose, nitric acid, and solvents to create the syrupy semisynthetic substance collodion that was the basis for Hyatt's experiments. Parkes had also noted the potential value of using camphor as a solvent for the material but didn't add it under heat and pressure as Hyatt did. His invention, Parkesine, was not quite as versatile or workable a material as celluloid, and he did not have Hyatt's marketing skills.

16 *The Victorian era:* Friedel, *Pioneer Plastic,* 28.

The noun plastic: The U.S. Patent Office didn't create a distinct category for plastics until 1903, and then it included oddities like compressed scraps of cork and leather, reflecting ongoing confusion over just what constituted plastic, according to Meikle, *American Plastic,* 5. According to Robert Malloy, chairman of the plastics engineering department at the University of Massachusetts at Lowell, the noun *plastic* first appeared in a dictionary in 1911. Author interview with Malloy, May 2010.

It shrugged off water: Robert Friedel, *A Material World: An Exhibition at the National Museum of American History* (Washington, DC: Smithsonian Institution, 1988), 41.

17 *One Colorado saloonkeeper:* Fenichell, *Plastic,* 41.

celluloid transcended the deficiencies: From Hyatt 1878 patent, quoted in Meikle, *American Plastic,* 16.

"Obviously none of the other materials": Hyatt patent, quoted in Keith Lauer and Julie Robinson, *Celluloid: Collector's Reference and Value Guide* (Paducah, KY: Collector Books, 1999), 102.

Celluloid could be rendered with the rich creamy hues: Though the product was popular with manufacturers, producing it wasn't an easy process. The material had to be made in a range of shades of white and yellow-white, then pressed into multiple laminations, and then sliced across the laminate grain in order to produce sheets that looked like ivory. Robert Friedel, e-mail correspondence with author, May 2010.

Celluloid made it possible: Meikle, *American Plastic,* 15.

"As petroleum came to the relief of the whale": Quoted in ibid., 12.

18 *Della, the young wife*: O. Henry, "The Gift of the Magi," in *The Best Short Stories of O. Henry* (New York: Modern Library, 1994), 1–7.

"few dollars invested in Celluloid": Quoted in Meikle, *American Plastic*, 15.

19 *Art critic John Ruskin described the thrill*: Quoted in ibid., 13.

The company urged its salesmen to emphasize: Friedel, *Pioneer Plastic*, 85–86. Celluloid dresser sets also filled a growing demand for more sanitary personal items, Friedel noted. Unlike silver, celluloid brushes and combs wouldn't tarnish, could be cleaned with soap and water, and were unaffected by moisture. And if made to look like ivory, they were white — the color of hygiene.

"with graining so delicate and true": "Ivory Py-ra-lin Toilet Ware de Luxe" (New York: The Arlington Co., 1917), quoted in Meikle, *American Plastic*, 18.

contemporary hairstyles often demanded: DuBois, *Plastics History*, 47.

Where once people had grown and prepared: Susan Strasser, *Satisfaction Guaranteed: The Making of the American Mass Market* (New York: Pantheon Books, 1989), 6.

20 *Celluloid was the first*: Meikle, *American Plastic*, 14.

Celluloid toothbrushes replaced: Susan Mossman, ed., *Early Plastics: Perspectives, 1850–1950* (London: Leicester University Press), 118.

billiards became an everyman's game: Author interview with Julie Robinson Robard, coauthor of *Celluloid: Collector's Reference and Value Guide,* in April 2010.

21 *In 1914, Irene Castle*: Glenn D. Kittler, *More Than Meets the Eye: The Foster Grant Story* (New York: Coronet Books, 1972), 1–2.

22 *A single machine equipped*: Meikle, *American Plastic*, 29.

DuPont . . . released photos: Ibid.

With the rise of mass-production plastics: Jen Cruse, *The Comb: Its History and Development* (London: Robert Hale, 2007). She writes: "The advent of injection-moulding machines, combined with the changing fashions in the early 1930s, killed off the idea of combs as decorative objects and the arts and skills of comb-making largely died out in Western countries" (page 54).

Combs were now stripped down: While celluloid eliminated the need for tortoiseshell to make combs, the tradition of making combs that look like tortoiseshell persists. Indeed, today tortoiseshell is just a part of the plastic palette; like the bright primary colors of Legos, tortoiseshell is a virtual guarantee that an object is plastic.

it took fifteen thousand beetles: Fenichell, *Plastic*, 87.

23 *it had a powerful identity of its own*: Friedel, e-mail correspondence with author, May 2010.

"stripped down as a Hemingway sentence": Fenichell, *Plastic*, 97.

"From the time that a man brushes his teeth": *Time*, September 22, 1924, quoted in Fenichell, *Plastic*, 97–98.

a shift in the development of new plastics: Friedel, *Pioneer Plastic*, 108.

When the first nylon stockings were introduced: Fenichell, *Plastic*, 147–49.

24 *thermoplastics quickly eclipsed:* Donald Rosato, Marlene Rosato, and Dominick Rosato, *Concise Encyclopedia of Plastics* (Boston: Kluwer Academic Publishers, 2000), 56; American Chemistry Council, "The Resin Review: 2008 Edition." The distinction is not always absolute; some polymers, such as polyethylene, can be formulated as either a thermoplastic or a thermoset. In addition, there are other, much smaller categories of polymers, including epoxies, silicone, and engineering plastics, polymers designed to meet high-performance demands.

25 *"Let us try to imagine":* V. E. Yarsley and E. G. Couzens, *Plastics* (Harmondsworth, UK: Penguin Books, 1941), 154–58.
Eager to conserve precious rubber: Author interview with Luther Hanson, curator of the U.S. Army Quartermaster Museum, April 2010.
Production of plastics leaped during the war: Meikle, *American Plastic,* 125. The role of Teflon in the atomic bomb comes from John Emsley, *Molecules at an Exhibition: The Science of Everyday Life* (Oxford: Oxford University Press, 1998), 133.

26 *DuPont had a whole division:* Mary Madison, "In a Plastic World," *New York Times,* August 22, 1943.
first National Plastics Exposition: "Host of New Uses in Plastics Shown," *New York Times,* April 23, 1946.
"Nothing can stop plastics": Ibid.
Plastics production expanded explosively: Meikle, *American Plastic,* 2. In the two decades following World War II, plastics grew at double-digit rates, faster than any major competing materials. Rosato et al., *Markets,* 2.
sell consumers on the virtue: One of the benefits often promoted by the industry was the ease with which plastics could be cleaned: all it took was the swipe of a damp cloth. A former editor of *Modern Plastics* recalled, "We used to write stories for many years—blah, blah, blah, and you can wipe it clean with a damp cloth—all the stories ended like that." Quoted in Meikle, *American Plastic,* 173.

27 *Hotels . . . began handing out complimentary combs:* Cruse, *The Comb,* 195–96.
we had an ever-growing ability to synthesize: Meikle, *American Plastic,* 2.
"You will have a greater chance to be yourself": Quoted in Thomas Hine, *Populuxe* (New York: Knopf, 1986), 4.

2. A Throne for the Common Man

28 *"Plastic as Plastic":* The exhibit was also good public relations for the chemical company underwriting its costs, Hooker Chemical Company, which was later found responsible for polluting Love Canal. Meikle, *American Plastic,* 1.
"the answer to an artist's dream": Hilton Kramer, " 'Plastic as Plastic': Divided Loyalties, Paradoxical Ambitions," *New York Times,* December 1, 1968.

29 *Herman Miller company reportedly spent:* Author interview with Peter Fiell, furniture historian, April 2008.
"chairs are extremely important": Author interview with Paola Antonelli, cura-

tor, Department of Architecture and Design, Museum of Modern Art, September 2008.

The oldest known chair: Florence de Dampierre, *Chairs: A History* (New York: Abrams Books, 2006), 17.

30 *The Shaker craftsmen:* Paul Rocheleau and June Sprigg, *Shaker Built: The Form and Function of Shaker Architecture* (New York: Monacelli Press, 1994), quoted in Paola Antonelli, "A Chair for the Common Man," unpublished manuscript.

The Thonet Model 14: Alice Rawsthorn, "No. 14: The Chair That Has Seated Millions," *International Herald Tribune,* November 7, 2008.

31 *"Every truly original idea":* George Nelson, *Chairs* (New York: Whitney Publications, 1953), 9, quoted in Charlotte Fiell and Peter Fiell, *1000 Chairs* (Cologne, Ger.: Taschen, 2000), 7.

The Greek root of the word: Meikle, *American Plastic,* 71.

As the French philosopher Roland Barthes: Roland Barthes, *Mythologies,* translated by Annette Lavers (New York: Hill and Wang, 1972), 97.

The makers of Bakelite: Meikle, *American Plastic,* 1.

32 *called* Plastikoptimismus: Quoted in ibid., 320.

Bakelite spoke "in the vernacular": Quoted in ibid., 108.

Karim Rashid: Rashid in e-mail interview with author, March 2008.

The Museum of Modern Art in 1956: Alison J. Clarke, *Tupperware: The Promise of Plastic in 1950s America* (Washington, DC: Smithsonian Institution Press, 1999), 36. In a 1947 article, the magazine *House Beautiful* called it "fine art for 39 cents." Ibid., 42.

33 *pseudo-wood cabinets:* The furniture industry's use of fake wood was so common by the late 1960s that the Federal Trade Commission warned the industry to stop using confusing and potentially actionable brand names such as Wonderwood and Miraclewood. "The only non-plastic furniture around ten years from now will be antiques," the merchandising director of Ward Furniture Company in Fort Smith, Arkansas, told *Modern Plastics* in 1968 in the article "The New Excitement in Furniture," *Modern Plastics* (January 1968): 89, 91.

Plastics' adaptability and glibness: Walter McQuade, "Encasement Lies in Wait for All of Us," *Architectural Forum* 127 (November 1967): 92, quoted in Meikle, *American Plastic,* 254.

people's unfortunate experiences with plastics: Meikle, *American Plastic,* 165–67.

The designer Charles Eames: Quoted in Fenichell, *Plastic,* 259. Even the plywood with which the Eameses often worked was a product of polymer technology; that sandwiching of wood layers hadn't been possible until plastics came along to laminate the veneers.

34 *he quit college to join the Danish resistance:* Panton's early years and career are discussed in a series of essays in Alexander von Vegesack and Mathias Remmele, eds., *Verner Panton: The Collected Works* (Weil am Rhein, Ger.: Vitra Design Museum, 2000). The essay by Mathias Remmele "All of a Piece: The Story of the Pan-

ton Chair" offers one of the few accounts of how Panton came to make his iconic chair.

He dressed only in blue: Author interview with Rolf Fehlbaum, CEO of Vitra, September 2008.

"should now use these materials": Panton, quoted in von Vegesack and Remmele, *Verner Panton,* 23.

35 *"we have not bothered about anything":* Poul Henningsen, quoted in ibid., 71.

At a design fair in 1959: Ibid., 28.

"It is at most a sculpture": The comment was reportedly made by George Nelson, design director of Herman Miller. See Remmele, "All of a Piece," in ibid., 78.

Panton didn't invent that form: Fiell and Fiell, *1000 Chairs;* author interview with Alexander von Vegesack, director of Vitra Design Museum, May 2008.

his vision brought the form into the synthetic age: Actually Panton executed a version of the S chair in molded plywood, but it didn't achieve the energetic feel or flow of the chair made in plastic. Nor could it be mass-produced. Von Vegesack and Remmele, *Verner Panton,* 76–77.

36 *Mid-twentieth-century designers:* The effort started in 1946 in Canada, when a pair of designers created a prototype fiberglass chair. Two years later, the Eameses began making their famed bucket seats of fiberglass. But neither of these was all of a single material or amenable to easy mass production. Fiberglass cannot be injection molded. Instead, the plastic-resin-and-fibers mixture that makes up fiberglass is poured into a mold and then sets; for years, that was a mostly non-mechanized, labor-intensive process.

Saarinen wanted seat and pedestal: Quoted in Fenichell, *Plastic,* 259.

He told colleagues: Quoted in Meikle, *American Plastic,* 204. Others working to develop plastic chairs were similarly stymied. In 1962, British designer Robin Day devised a chair with a body made of inexpensive polypropylene. Working with the Shell Corporation (which held the patent on the plastic), he was able to master the molding technology to produce the seat in a single piece. But the metal legs still had to be attached separately. Nonetheless, the Polyprop was a stunning success, becoming the classic institutional plastic chair still found in classrooms and waiting rooms around the world. Some fifteen million have been sold to date, and in 2008 the British government issued a stamp celebrating the chair (as part of a series honoring the best of British design). A few years after the Polyprop, the German architects Helmut and Alfred Batzner designed a chair made of a fiberglass-like material that could be molded in a single piece, but again it wasn't readily mass-produced, and the legs were made separately. Author interview with Peter Fiell, and Fiell and Fiell, *1000 Chairs.*

37 *the Italian company Kartell:* Indeed, the Italians in general led the way in the development of plastic furniture, thanks in part to close cooperation among Italian big business, small workshops, and progressive designers, according to Karl Mang, *History of Modern Furniture* (New York: Harry Abrams, 1978), 160.

The Castellis recognized: Catalog of Kartell Museo.

"the odor of the refinery": Meikle, *American Plastic,* 226.

38 *even Kartell had trouble:* Author interview with Peter Fiell.

The company's owner wasn't wild: Author interview with Rolf Fehlbaum.

The chair proved more challenging: Panton, Fehlbaum, and Vitra engineer Manfred Dieboldt wanted to find a plastic that could be machine-molded and that wouldn't require extensive hand-finishing afterward. In the first stab at making the chair, the company used a type of polyester reinforced with fiberglass, which required a lot of costly, time-consuming manual labor. The few prototypes they made were an instant sensation in the design press, but Panton was unhappy with the weight of the chair, the uneven finish, and the high price. See von Vegesack and Remmele, *Verner Panton,* 85.

In 1968, they found the perfect plastic: Though Panton had broken through the polymer ceiling in chair manufacture, he wasn't entirely satisfied with the Baydur chair, which came out of the mold still looking and feeling rough. It had to be spackled, sanded, and painted, which drove up the price. Panton wanted to make the chair affordable. Two years later, he and his partners came upon a new plastic they thought would work even better: a polystyrene called Luran S, made by BASF, which came in dyed granules and needed no additional processing once out of the mold. They began making the chair from Luran S. But after several years, it became apparent the material was less hardy than initially thought. Consumers complained that their chairs cracked, and even broke. Those complaints, coupled with changing tastes, led to a significant decline in sales; in 1979 the company stopped making the chair. Panton once again had to go in search of a manufacturer who believed in the beauty and value of the design. Eventually, Vitra, the successor to the company that had first produced the chair, picked it up again, and since 1990 it has been making and selling the chair. Information from von Vegesack and Remmele, *Verner Panton,* 84–86, and author interviews with Fehlbaum and von Vegesack and with Manfred Dieboldt, former engineer for Vitra, December 2009.

39 *"It embodies the enthusiasm":* von Vegesack and Remmele, *Verner Panton,* 94.

One magazine featured a model posing: From a biographical essay on Panton posted on the website of London's Design Museum, http://designmuseum.org/design/verner-panton.

40 *There are hundreds of millions:* Alice Rawsthorn, "Celebrating the Everychair of Chairs," *International Herald Tribune,* February 4, 2007; Mariana Gosnell, "Everybody Take a Seat," *Smithsonian* magazine, July 1, 2004. Both authors have written perceptively about the ubiquitous but essentially invisible monobloc.

one conspiracy-minded blogger: Debate over the significance of the chair in the video of Berg's killing took place on www.functionalfate.org, a website devoted to the monobloc chair.

They've been spotted: Jens Thiel, "Sit Down and Be Counted," *ArtReview* (April 2006): 58–61.

41 *far beyond the rarefied realm of design:* Author interview with Jens Thiel, creator of www.functionalfate.org, March 2008.

the raw plastic could be had for less than: Author interview with Stephen Green-berg, Canadian Mercantile Trading Corporation, June 2008.

Though monobloc chairs are cheap: Cost of an injection press from Gosnell, "Take a Seat"; cost of molds from author interview with George Lemieux, KCA Plastics Consultants, April 2008.

42 *intense competition eventually winnowed:* The playing field is less cluttered, but plastic furniture still remains a difficult business. For instance, Bill Adams, the founder of Adams Manufacturing, doesn't make a penny on his basic model. He considers the chair a loss leader, as do the retailers that sell it — they all but give it away in order to get people into the store.

Adams had no regrets: Author interview with Bill Adams, the founder of Adams Manufacturing, March 2008.

45 *"I have six monoblocs":* In an age when irony is as abundant as plastic, it is perhaps inevitable that even the profoundly unhip monobloc would eventually acquire some cachet — if not strictly as furniture, then at least as cultural commentary. Spanish designer Marti Guixé graffitied plastic chairs with slogans such as *Stop Discrimination of Cheap Furniture* and *Respect Cheap Furniture* and then sold them as limited editions (no doubt for more than the price of a six-pack). Other conceptual artists have shredded them, stacked them in huge piles, hung them from buildings, drilled them full of tiny holes, even remade them in wood.

there are an estimated one hundred manufacturers: Author interview with Thiel.

design of monoblocs is largely the result: Author interviews with Lemieux, Adams.

46 *a throne for the common man:* The phrase comes from Antonelli's unpublished essay "A Chair for the Common Man."

47 *Hank Stuever summed up the scorn:* Quoted in Gosnell, "Take a Seat."

"in the city of Basel": Fehlbaum's hometown is not unique. A number of cities, in-cluding Bratislava, Helsinki, Berne, Switzerland, and Los Gatos, California, have passed laws forbidding businesses from putting out the chairs. The Los Gatos ordinance was designed to protect the "quiet" character of it 1940s-era historic downtown.

48 *Ultradur:* It's a relative of the same plastic used to make soda bottles, polyethylene terephthalate, or PET.

Alice Rawsthorn praised: Alice Rawsthorn, "Konstantin Grcic's New Chair De-sign, the MYTO," *International Herald Tribune,* November 9, 2007.

49 *Starck's feelings about plastic:* Author interview with Philippe Starck, May 2008.

50 *"when we design a chair":* Quoted in Fiell and Fiell, *1000 Chairs,* 9.

3. Flitting Through Plasticville

53 *Since the flying discs were introduced:* Wham-O won't divulge sales figures; that's the estimate of a source familiar with the company.

Plastics constitute: The basic industry statistics come from the Society of the Plas-tics Industry, "Fast Facts on Plastics," accessed on the SPI website, http://www

.plasticsindustry.org/Press/content.cfm?ItemNumber=798&navItemNumber=
1323. According to the SPI, there are nearly 18,500 plastics-related facilities in the
United States, with the largest number based in Texas and California.

Waisblum, who at that point oversaw: He left Wham-O in 2009 after new owners
bought the company.

54 *Morrison joined his girlfriend Lucille's family:* Morrison told the story in his 2006
book *Flat Flip Flies Straight! True Origins of the Frisbee* (Wethersfield, CT: Worm-
hole Publishers, 2006), which he cowrote with Frisbee collector Phil Kennedy.
The title refers to the directions originally written by Morrison's wife, Lucille, and
which are still pressed into the underside of every Frisbee: *Flat flip flies straight,
Tilted flip curves — Experiment! Play catch — Invent games.*

55 *another redesign:* Tim Walsh, *Wham-O Super-Book: Celebrating Sixty Years Inside
the Fun Factory* (San Francisco: Chronicle Books, 2008), 77.
One day, when Morrison was demonstrating: Morrison and Kennedy, *Flat Flip
Flies Straight!*, 106.

56 *Wham-O's early catalog:* Walsh, *Wham-O Super-Book*, 30–34.
"You couldn't buy those things": Ibid., 17.
Though there had been plastic toys: Celluloid rattles and kewpie dolls were popu-
lar, and in the 1930s celluloid acetate was used to make toy musical instruments.
"where childish hands find nothing to break": Yarsley and Couzens, *Plastics,* 154.
During the peak years of the baby boom: Donovan Hohn, "Moby-Duck: Or, The
Synthetic Wilderness of Childhood," *Harper's* magazine (January 2007): 57. To-
day, three billion toys are sold annually in the United States, according to Michael
Luzon, "No Toying Around," *Plastics News,* December 22, 2008.
an ever-increasing number of those toys: Hiram McCann, "Doubling, Tripling, Ex-
panding: That's Plastics," *Monsanto* magazine (October 1947): 5.
Today, plastics are a given: Author interview with Robert von Goeben, cofounder
of Green Toys, September 2007.
Fleshy vinyl permitted: "Trends in Toys," *Modern Plastics* (June 1952): 71.
Silly Putty: Fenichell, *Plastic,* 260–62.

57 *To promote its house brand:* Bill Hanlon, *Plastic Toys: Dimestore Dreams of the '40s
and '50s* (Atgen, PA: Schiffer Publishing, Inc., 1993), 10–14.
eight different chemical companies quickly built factories: By that time, the original
inventors of polyethylene, DuPont and Britain's Imperial Chemical Industries,
had lost control of their invention in an antitrust action. That opened the doors
for other chemical companies to start producing it. Meikle, *American Plastic,*
189–90.
pop beads: Ibid., 190.
Such boom-bust cycles: The industry's growth has slowed in recent years. For instance, as *Plastics News* reported, the total compounded sales growth for plastic
sales between 1973 and 2007 was 4.2 percent, but in the final six years of that period, the rate of growth was half that amount. The slowdown at the end of that
thirty-four-year-long stretch reflects how the market cooled "after major conver-

sions from materials like wood and metal took place in the 1970s and 1980s . . . it's also a sign of how production moved from North America to parts of the world with lower labor and production costs." Frank Esposito, "Resin Market Slows in North America," *Plastic News*, September 10, 2007.

the ping-ponging relationship: Meikle, *American Plastic*, 190; Walsh, *Wham-O Super-Book*, 62–69. At the fad's height, Wham-O was making twenty thousand hoops a week.

58 *"We damn near went broke":* Walsh, *Wham-O Super-Book*, 69.

Melin and Knerr rechristened Morrison's baby: Ibid., 78–79. Morrison groused about the name, complaining it didn't describe anything. But for all his grumbling, he recognized, his deal with Wham-O made him wealthy, and for many years he continued to work with the company, promoting flying discs. By the time Morrison died, in 2010 at the age of ninety, Wham-O had changed hands repeatedly, and he had little contact with it anymore.

Headrick redesigned the Frisbee: Ibid., 191.

59 *Standard Oil was the first to figure out:* Gerald Markowitz and David Rosner, *Deceit and Denial: The Deadly Politics of Industrial Pollution* (Berkeley: University of California Press, 2002), 235.

60 *production of plastics consumes:* Anthony Andrady and Mike A. Neal, "Applications and Societal Benefits of Plastics," *Philosophical Transactions of the Royal Society B* 364 (June 2009): 1980; Anthony Andrady et al., "Environmental Issues Related to the Plastics Industry: Global Concerns," in Andrady, ed., *Plastics and the Environment* (Hoboken, NJ: John Wiley and Sons, 2003), 38. The American Chemistry Council states that plastics production accounts for about 4 to 5 percent of natural gas consumed in the United States annually and about 3 percent of oil.

an industry based on waste: Barry Commoner, introduction to Geiser, *Materials Matter*. Also, author interview with Ken Geiser, director of the Lowell Center for Sustainable Production, January 2010.

With the rise of integrated: Meikle, *American Plastic*, 82–83.

Polyethylene was discovered: Fenichell, *Plastic*, 200–202; ICI chemist quoted in Meikle, *American Plastic*, 189.

61 *British took advantage of that dielectric quality:* Fenichell, *Plastic*, 202.

"stiffer than steel": Colin Richards, "Polyethylene, a Phenomenon," *Plastiquarian* 40 (October 2008): 14.

polyethylene was the first plastic: Society for the Plastics Industry, "Definition of Resins-Polyethylene," accessed at http://www.plasticsindustry.org/AboutPlastics/content.cfm?ItemNumber=1400&navItemNumber=1128.

62 *we encounter many other kinds:* Richard Thompson et al., "Our Plastic Age," *Philosophical Transactions of the Royal Society B* 364 (2009): 1973. The *Concise Encyclopedia of Plastics*, page 57, states there are about seventeen thousand different varieties of plastic, and other experts offer higher estimates, in the range of thirty thousand.

the five basic families of commodity plastics: American Chemistry Council, *The Resin Review 2008*, 16–17. U.S. polymer production peaked at 115 billion pounds in 2007. But with the recession that followed, resin production fell below 100 billion pounds for the first time in a decade as demand from various end markets shrank, according to the American Chemistry Council.

No significant new plastics have been introduced: Author interviews with industry consultant Glenn Beall, Glenn Beall Plastics, Ltd., March 2008; and David Durand, senior consultant, Townsend Solutions, March 2008. To illustrate the challenges of making entirely new polymers, Beall described General Electric's fifteen-year, $50 million push to make Ultem, or polyetherimide, a plastic designed to withstand very high temperatures. By the time the plastic was ready for the market, the patents had just about run out. Kevlar, the material used in bulletproof vests, cost DuPont $500 million before its 1982 launch. Emsley, *Molecules at an Exhibition*, 143.

For decades, it's constituted about a third: American Chemistry Council, *Resin Review 2008*, 25–31.

according to calculations by Skidmore College: Author interview with Raymond Giguere, June 2008. Giguere did his calculations in 2006 and assumed that the average weight (mass) of an American was 150 pounds and that the U.S. population was 300 million, which results in a total of 45 billion pounds. The U.S. produced about 39 billion pounds of polyethylene in 2006. Production was down slightly in 2009, to 26 billion pounds, but the population had certainly increased.

63 *Dow arrived here in 1940:* Dow had gotten into that business accidentally. In the process of harvesting minerals from brine, the company accumulated byproducts, such as ethyl chloride. Company chemists began investigating ways to use them; one of the products proposed was ethyl cellulose, a semisynthetic polymer made with wood pulp, which Dow began marketing in 1935 under the name Ethocel. Likewise, the company got into making polystyrene in the late 1930s as a way to deal with excess supplies of ethylene. Ethylene could be reacted with benzene, another processing waste product, to form ethyl benzene, which in turn could be made into styrene, the base ingredient for polystyrene, the plastic Dow sold under the trademarked name Styron. Jack Doyle, *Trespass Against Us: Dow Chemical and the Toxic Century* (Monroe, ME: Common Courage Press, 2004), 146–47.

contaminating the ground water: Dina Cappiello and Dan Feldstein, "In Harm's Way: A Special Report," *Houston Chronicle*, January 20, 2005, and author interview with Sharron Stewart, longtime environmental activist in the Freeport area, February 2009.

64 *It pays more than $125 million in state and local taxes:* Author interview with Tracie Copeland, Dow Chemical, February 2009.

65 *about 70 percent of plastics:* Author interview with Howard Rappaport, global business director, Chemical Market Associates, Inc., February 2009. Most crackers are able to process only one or the other.

two carbons can bond to form the gas ethylene: Ethylene is the largest-volume chemical made, and half of all ethylene produced is used to make polyethylene.

66 *"Dow has come a long way":* Author interview with Charles Singletary, business manager, Local 564, International Union of Operating Engineers, February 2009.

68 *These pellets, also known as nurdles:* Some plastics, such as polyvinyl chloride, are shipped in the form of powder rather than as pellets.

69 *a seismic change is under way:* Steve Toloken, "Industry Shifts Toward Asia Continues," *Plastics News,* March 16, 2009; author interview with Rappaport; presentation by Rappaport, "Economy, Energy, Feedstocks, Polymers and Markets," March 2009.

the Middle East's share: Author interview with Rappaport.

the Saudis are trying: Ibid. The country is building two complexes along the Red Sea coast to lure investors who want to set up plastics-production facilities. As Rappaport put it, the Saudi government is essentially telling product manufacturers, " 'We can supply you with the pellets here and you can make your finished goods, and we'll ship finished goods around rather than pellets.' Which is what they do in China. It's the China model, except the Middle East has a lower raw-material cost."

A wide gap still exists: Li Shen, Juliane Haufe, and Martin K. Patel, "Product Overview and Market Projection of Emerging Bioplastics," a report commissioned by European Polysaccharide Network of Excellence and the European Bioplastics Council, November 2009, 7.

plastics production to swell: Ibid., 8.

70 *"an 'ATM unit' ":* Author e-mail correspondence with Danny Grossman, December 2009.

71 *Every mold maker in Southern California:* Author interview with Clare Goldsberry, contributing editor, *Injection Molding Magazine,* June 2008.

72 *Wham-O stayed put until:* Mattel owned Wham-O briefly, from 1994 to 1997, and then sold it to a group of American investors, who next sold it to Hong Kong–based investors. In 2009, an American company, Manufacturing Marvel, bought it.

a place that's been described: James Fallows, *Postcards from Tomorrow Square: Reports from China* (New York: Vintage Books, 2009), 66.

"the heart pumping China's emergence": Robert Marks, "Robert Marks on the Pearl River Delta," *Environmental History* 9 (2004): 296.

As many as fifty thousand factories: Steve Toloken, "Ecologists Seeking to Clean Up China's PRD Zone," *Plastics News,* December 31, 2008.

twice as many people: Steve Toloken, Guangzhou-based reporter for *Plastics News,* in e-mail correspondence with author, July 2010.

foreign investment at the incredible rate: Marks, "Pearl River Delta," 296–97.

Shipping containers left: Fallows, *Postcards,* 6.

If the region were a country: Michael J. Enright et al., *The Greater Pearl River Delta* (Hong Kong: Invest Hong Kong, 2007), 1. The estimate is based on 2005 data.

By the 1990s, the silkworms: Marks, "Pearl River Delta," 297.

73 *Guangdong has been a locus for international trade:* Sun Qunyang, Larry Qiu, and Li Jie, "The Pearl River Delta: A World Workshop," in Kevin H. Zhang, ed., *China as the World Factory* (London: Routledge, 2006).

Hong Kong had a strong plastics-processing industry: Hong Kong Government Industry Department, "Hong Kong's Manufacturing Industries," December 1996. Author interviews with L. T. Lam, Forward Winsome Industries, Tony Lau, Canfat Manufacturing, Dennis Wong, March 2009. A few manufacturers tried earlier to set up shop in the mainland. Lam, founder of Winsome Industries, claimed to have been one of the very first plastic-toy makers in Guangdong. He opened a factory there in the 1940s but then had to retreat back to Hong Kong when the Communists took power in 1949. When we met in Hong Kong, he showed me what he claimed was one of the first plastic toys ever produced in Asia: a whistle with a bird in a little round cage on the top.

74 *Enticed through Deng's open door:* According to author Leslie Chang, the first mainland factory was the Taiping Handbag Factory of Hong Kong, which opened in Dongguan in 1978 and made one million in Hong Kong dollars in its first year. "The factory processed material from Hong Kong into finished goods, which were shipped back to Hong Kong to be sold to the world. It established the model for thousands of factories to follow." Leslie Chang, *Factory Girls: From Village to City in a Changing China* (New York: Spiegel and Grau, 2009), 29.

78 *the average worker at Mattel's plant:* Jonathan Dee, "A Toy Maker's Conscience," *New York Times Magazine,* December 25, 2007.

China Labor Watch reported: China Labor Watch, "Investigation on Toy Suppliers in China: Workers Are Still Suffering," August 2007.

79 *products are destined to go overseas:* China's domestic toy market is still minuscule; retail toy sales totaled $603 million in 2006, compared to the more than $20 billion Americans spend on toys. But this is starting to change with the rise of a Chinese middle class. Those parents who now can afford it often shop for foreign-brand toys, such as Legos from Denmark or Transformers made by Japan-based Bandai. Even if the toys were actually made in China, Chinese parents assume the foreign brand name means they will be safer and less likely to contain hazardous materials such as lead paint. Elaine Kurtenbach, "Chinese Kids Get Foreign-Brand Toys," Associated Press, December 14, 2007.

epidemic of toy recalls: See, for instance, Michael Lauzon, "Chinese Toy Recalls May Be Boon to U.S.," *Plastics News,* December 17, 2007.

more than five thousand toy companies: Steve Toloken, "Safety Concerns Cost Chinese Toy-Makers," *Plastics News,* January 27, 2009.

4. *"Humans Are Just a Little Plastic Now"*

82 *Dutch physician Willem Kolff:* Fenichell, *Plastic,* 329–30. Kolff later mentored Robert Jarvik, the inventor of the first successful artificial heart, which surgeon

William DeVries said snapped into place "just like closing Tupperware" when he implanted the first one in 1982.

With plastics, hospitals: "Boom in Single-Use Markets," *Modern Plastics* (March 1969): 60–62.

medicine is a small end market: A 2010 report by Global Industry Analysts suggests medical plastics will consume about 10 billion pounds of all polymers produced globally by 2015. Global plastics production is approximately 570 billion pounds. But many major producers of medical supplies are based in the United States, meaning medicine is a bigger end market for the domestic industry. In an interview with the author in August of 2009, Ken Pawlak, industry consultant and author of a forthcoming book on medical plastics, estimated medicine accounted for 10 percent of plastic consumption. Consumption figures for other end markets come from the American Chemistry Council, *The Resin Review* 2007.

a strong, recession-proof market: It's also growing faster than many other end markets. See Mike Verespej, "Medical Faring Better Than Many Markets," *Plastics News,* January 13, 2009; "Medical Suppliers Optimistic Their Market Will Remain Strong," *Plastics News,* July 1, 2010.

enormous PR value: One of the biggest exhibits in the now-closed National Plastics Museum in Leominster, Massachusetts, was devoted to medical applications of plastics, including the Spare Parts Body Shop, a display of various plastic-based prosthetics. The neonatal incubator was featured in the American Chemistry Council's proplastics campaign Essential2.

84 *they play havoc:* For good overviews of the literature on endocrine disrupters in plastics, see John Wargo et al., "Plastics That May Be Harmful to Children and Reproductive Health," Environment and Human Health, Inc., North Haven, CT, 2008. See also Center for Evaluation of Risks to Human Reproduction (CERHR), "NTP-CERHR Monograph on the Potential Human Reproductive and Developmental Effects of Di(2-Ethylhexyl) Phthalate (DEHP)," November 2006; National Research Council, *Phthalates and Cumulative Risk Assessment: The Task Ahead* (Washington, DC: National Academies Press, 2008).

concentrations we never considered worrisome: Bisphenol A has been found to have action at the parts-per-trillion range. See Wargo, "Plastics That May Be Harmful," 22; Frederick vom Saal et al., "Chapel Hill Bisphenol A Expert Panel Consensus Statement: Integration of Mechanisms, Effects in Animals and Potential to Impact Human Health at Current Levels of Exposure," *Reproductive Toxicology* 24 (August–September 2007): 131–38.

85 *Like many surgeons:* John R. Brooks, "Carl W. Walter, MD: Surgeon, Inventor, and Industrialist," *American Journal of Surgery* 148 (November 1984): 555–58.

university trustees who considered it "unethical and immoral": Quoted in research summary e-mail to author from Hank Grasso, DeWitt Stetten Jr. Museum of Medical Research, August 2009.

Blood banks of that era: Robert Ausman and David Bellamy Jr., "Problems and

Resolutions in the Development of the Flexible Plastic Blood Container," *American Journal of Surgery* 148 (November 1984): 559–61.

PVC is a unique polymer: PVC was first created in 1872 but not commercially produced until 1920. More than half the molecule (about 57 percent by weight) is composed of chlorine. Andrady, "Applications and Societal Benefits," 1978.

86 *it can be "converted into an almost limitless":* Vinyl Institute website, www.vinylinfo .org.

Such versatility has made PVC: Vinyl is one of the largest-volume plastics produced in North America. More than fifteen billion pounds were manufactured in the United States in 2006, according to the Vinyl Institute.

a frequent choice for makers of medical devices: PVC constitutes about a quarter of the plastics used in medical devices. Joel Tickner et al., "The Use of Di-2-Ethylhexyl Phthalate in PVC Medical Devices: Exposure, Toxicity, and Alternatives," Lowell Center for Sustainable Production (1999): 9.

plasticized PVC: Prior to the 1930s, castor oil and later camphor were the common chemicals used to soften hard plastics.

Phthalates have become so ubiquitous: Over 470 million pounds of phthalates are produced each year, according to the EPA, "Phthalates Action Plan," December 12, 2009. Accessed at http://www.epa.gov/oppt/existingchemicals/pubs/ actionplans/phthalates_ap_2009_1230_final.pdf. DEHP accounts for slightly more than half of phthalates produced. In 2002 manufacturers produced about 260 million pounds of the chemical, according to the Agency for Toxic Substances and Disease Registry (ATSDR), "Toxicological Profile for Di(2-ethylhexyl)phthalate (DEHP)," Production, Import, Use, and Disposal, 2002. Globally, about a billion pounds of phthalates are produced annually, according to the Our Stolen Future website, http://www.ourstolenfuture.org/newscience/ oncompounds/phthalates/phthalates.htm#.

They're used as plasticizers: Ted Schettler, "Human Exposure to Phthalates Via Consumer Products," *International Journal of Andrology* 29 (February 2006): 134.

you'll find phthalates in other types: Ibid.; Wargo et al., "Plastics That May Be Harmful."

They're even used in the time-release coating: Schettler, "Human Exposure"; S. Hernandez-Diaz et al., "Medications as a Potential Source of Exposure to Phthalates in the U.S. Population," *Environmental Health Perspectives* 117 (February 2009): 185–89.

There are about twenty-five: Use of various phthalates depends on their molecular weight, that is, the mass of the molecule. Higher-weight phthalates, such as DEHP, di-isononyl phthalate (DINP), and di-isodecyl phthalate (DIDP), are produced in the highest volume and used in construction materials, clothing, and furnishings. Lower-weight phthalates, such as dibutyl phthalate (DBP), diethyl phthalate (DEP), and dimethyl phthalate (DMP), tend to be used as solvents and in adhesives, waxes, inks, cosmetics, insecticides, and pharmaceuticals. See Schettler, "Human Exposure," 134.

DEHP is one of the most popular: About a quarter of the DEHP manufactured is used in medical devices. ATSDR, "Toxicological Profile."

87 *To persuade colleagues:* Douglas M. Surgeonor, "Reflections on Blood Transfusion," *American Journal of Surgery* 148 (November 1984): 563.

The new technology revolutionized: Author interview with Gary Moroff, American Red Cross, November 2009.

The U.S. Army employed: Author interview with Pawlak. In 1959, Walter sold Fenwal, the company he founded to market his blood-bag system, to medical supply giant Baxter Healthcare.

One of PVC's big selling points: "Why Doctors Are Using More Plastics," *Modern Plastics* (October 1957): 87.

88 *Doctors at B. F. Goodrich's:* Markowitz and Rosner, *Deceit and Denial,* 173–75.

European researchers found evidence: Ibid., 171.

"a plastic coffin": Quoted in ibid., 192.

Industry howled: Ibid., 223.

another line of research: This line of inquiry actually went back to the 1940s, with scattered reports that some substance migrating out of various plastic films could induce tumors in rats. In the mid-1950s, for instance, a group of Columbia University researchers happened onto disturbing findings about the newly introduced plastic films, such as Saran Wrap, which Dow promoted as "the film of one hundred and one uses." In a use that surely wasn't on Dow's list, the researchers had wrapped lab rats' kidneys in plastic film for a study on hypertension drugs. To their surprise, several years later, they found that seven of the rats had developed malignant tumors at the sites where their kidneys had been wrapped. In later studies, they found that tumors sprouted in high rates in rats exposed to a number of different plastics, including Saran Wrap (made of polyvinylidene chloride, a cousin of PVC), PVC, polyethylene, Dacron, cellophane, and Teflon. It wasn't clear to the researchers what was causing the tumors or whether the rats' disease signified a risk for human health. In the 1970s, the FDA's concern that vinyl chloride could leach out of PVC led it to turn down Monsanto's request to make PVC bottles for liquor. Sarah Vogel, "The Politics of Plastic: The Social, Economic and Scientific History of Bisphenol A," PhD dissertation, Columbia University, 2008.

Johns Hopkins University toxicologists: The story of their discovery comes largely from author interviews with Rudolph Jaeger, September 2009, and Robert Rubin, October 2009. See also R. J. Jaeger and R. J. Rubin, "Plasticizers from Plastic Devices Extraction, Metabolism, and Accumulation by Biological Systems," *Science* 170 (October 23, 1970): 460–62; R. J. Jaeger and R. J. Rubin, "Contamination of Blood Stored in Plastic Packs," *Lancet* 2 (July 18, 1970): 151; R. J. Jaeger and R. J. Rubin, "Some Pharmacologic and Toxicologic Effects of Di-2-Ethylhexyl Phthalate (DHP) and Other Plasticizers," *Environmental Health Perspectives* (January 3, 1973): 53–59.

89 *bags could be as much as 40 percent:* Tickner et al., "Use of Di-2-Ethylhexyl Phthalate."

The additive is not atomically bonded: Indeed, leaching is a virtual certainty given the architecture of plasticized PVC, according to toxicologist Bruce LaBelle. Author interview with LaBelle, California Department of Toxic Substances Control, September 2009. Over time, the normal forces of atomic attraction pull the long PVC molecules together, which eventually squishes the DEHP out. That process is creating a crisis in the modern art world, where conservators are struggling to find ways to contend with plastic artworks that are weeping plasticizers or off-gassing unpleasant smells. Sam Kean, "Does Plastic Art Last Forever?" *Slate* magazine, July 1, 2009.

"Humans are just a little plastic": Victor Cohn, "Plastics Residues Found in Bloodstreams," *Washington Post,* January 18, 1972.

90 *After taking a hard look at DEHP:* As one 1978 review of the literature put it: "there is no evidence of toxicity from the use of PVC plasticized plastics in medical practice. The major components of plasticized PVC have been examined over a span of years and each passing year sees a confirmation of the lack of toxicity . . . Considering all the factors of cost, convenience, and safety, it appears that plasticized PVC containers continue to have a valuable place in medical practice." W. L. Guess, "Safety Evaluation of Medical Plastics," *Clinical Toxicology* 12 (1978): 77–95. See also Naomi Luban et al., "I Want to Say One Word to You—Just One Word—'Plastics,'" *Transfusion* 46 (April 2006): 503–6.

poison is "a quantitative": Ernest Hodgson and Patricia Levi, *A Textbook of Modern Toxicology* (New York: Elsevier, 1987), 2.

The dose makes the poison: Pete Myers and Wendy Hessler, "Does the Dose Make the Poison?" *Environmental Health News,* April 30, 2007.

Theo Colborn began developing a different theory: The story of Colborn's work and evolving understanding of endocrine disrupters is told in Theo Colborn, Dianne Dumanoski, and John Peterson Myers, *Our Stolen Future: Are We Threatening Our Fertility, Intelligence, and Survival? A Scientific Detective Story* (New York: Dutton, 1996), 12. See also Vogel, "Politics of Plastic"; Gay Daly, "Bad Chemistry," *OnEarth* (Winter 2006).

91 *as one reporter observed:* Daly, "Bad Chemistry."

"something important was lurking": Colborn et al., *Our Stolen Future,* 12.

92 *"hand-me-down poisons":* Ibid., 26.

the drug DES: Recent animal studies have suggested third-generation effects among DES exposed mice, though the risks to human DES grandchildren are not yet clear. Wargo et al., "Plastics That May Be Harmful," 8.

she organized a meeting: Daly, "Bad Chemistry"; Vogel, "The Politics of Plastic," 238–40.

"I was scared to death!": Daly, "Bad Chemistry."

93 *They dubbed it "endocrine disruption":* In hindsight, said Ted Schettler, a leading

researcher in the field, the choice of phrase was "a little unfortunate." It focused attention on the hormonal pathways affected by synthetic chemicals. "But there are many other signaling pathways that are important for normal physical development or function." More recent research has begun looking at the effects of chemicals on neurochemical messengers in the brain, among others. Author interview with Schettler, science director, Science and Environmental Health Network, October 2009.

Its hallmarks included: Vogel, "The Politics of Plastic," 244.

The first Wingspread conference: Colborn et al., *Our Stolen Future,* 253.

Today, the number may be: Japanese regulators have identified seventy endocrine disrupters (Daly, "Bad Chemistry"); the figure of one thousand comes from John Wargo, "Pervasive Plastics: Why the U.S. Needs New and Tighter Controls," *Yale Environment 360* (November 16, 2009).

by mimicking natural hormones: Colborn et al., *Our Stolen Future,* 72.

bisphenol A: The chemical is also present in PVC.

the bonds holding these long molecules: Author interviews with Fred vom Saal, University of Missouri, Columbia, October 2007, and Bruce LaBelle. See also Vogel, "Politics of Plastic"; Frederick vom Saal et al., "An Extensive New Literature Concerning Low-Dose Effects of Bisphenol A Shows the Need for a New Risk Assessment," *Environmental Health Perspectives* 113 (August 2005): 926–33 and Wargo et al., "Plastics That May Be Harmful."

two possible ways to cause static: Wargo et al., "Plastics That May Be Harmful," 22.

it makes sense: Author interview with vom Saal.

94 *Hormones are produced:* Colborn et al., *Our Stolen Future,* 32.

a known neurotoxin: Long-term occupational exposure to the chemical may have subtle neurological effects, and a recent report issued by the American Cancer Society, the National Institute for Occupational Safety and Health, the National Institute of Environmental Health Sciences, and the National Cancer Institute included styrene among twenty potential carcinogens deserving of more investigation. Reuters, "Report Targets Twenty Possible Causes of Cancer," July 15, 2010.

95 *thirty-seven-billion-dollar global market:* BCC Research, "Plastic Additives: The Global Market," June 2009. Synopsis accessed at http://www.bccresearch.com/report/PLS022B.html.

the recent experience of German researchers: Author interview with Martin Wagner, Department of Aquatic Ecotoxicology, Goethe University, Frankfurt am Main, May 2009. See also M. Wagner et al., "Endocrine Disruptors in Bottled Mineral Water: Total Estrogenic Burden and Migration from Plastic Bottles," *Environmental Science Pollution Research International* 16 (May 2009): 278–86. Italian researchers reported similar results: B. Pinto et al., "Screening of Estrogen-Like Activity of Mineral Water Stored in PET Bottles," *International Journal of Hygiene and Environmental Health* 212 (March 2009): 228–32. Leaching of chemicals from PET isn't entirely surprising. Typically, a tiny fraction of the plastic—about 1 percent—consists of molecules that never completely polymerized.

These shortened daisy chains, known as oligomers, might be only a few units long. Because they're smaller than polymer molecules, oligomers can sneak out of the plastic matrix, carrying with them any chemical additives or manufacturing residues.

one possibility is antimony: Ted Schettler in e-mail to author, November 2009. George Bittner, a neuroscientist at the University of Texas, contends that the reports of hormonal activity in polycarbonate, vinyl, and PET are "the tip of the iceberg." Bittner claims to have tested hundreds of common plastics and additives in cell studies and has yet to find any that don't show the capacity to mimic hormones. However, as of mid-2010, his work had not been published in any peer-reviewed journals.

there's no way for consumers to know: That's true even with cosmetics, which are subject to strict labeling requirements. One study analyzed seventy-two different cosmetic and personal-care products; phthalates weren't listed on the labels of any but were found to be present in fifty-two of the products. Schettler, "Human Exposure," 137.

96 *they are odorless:* In 2008, the Center for Health, Environment, and Justice tried to pin down the source of that chemical smell by analyzing new vinyl shower curtains. The analysis, done by independent labs, suggested the smell was not produced by a single chemical but by a mix of dozens. Indeed, the researchers found new vinyl shower curtains contained as many as 108 different volatile organic chemicals, including DEHP and another phthalate, DINP. "Some of these chemicals cause developmental damage as well as damage to the liver and central nervous, respiratory, and reproductive systems." Stephen Lester et al., "Volatile Vinyl: The New Shower Curtain's Chemical Smell," Center for Health, Environment, and Justice, June 2008.

nonmedical deployment: Schettler, "Human Exposure"; Wargo et al., "Plastics That May Be Harmful"; CERHR, "Monograph." The link to flip-flops was in a recent report published by Swedish Society for Nature Conservation, "Chemicals Up Close: Plastic Shoes from All Over the World," 2009, which found phthalates in seventeen of twenty-seven shoes tested. For a review of studies looking at exposure through food, see Jane Muncke, "Exposure to Endocrine Disrupting Compounds Via the Food Chain: Is Packaging a Relevant Source?" *Science of the Total Environment* 407 (August 2009): 4549–59. Much of the food-related research has been done in Europe, and it's not clear how well it applies to American markets.

97 *Once the compound enters the bloodstream:* Author interviews with Shanna Swan, University of Rochester, October 2007 and October 2009. Also Mark Schapiro, *Exposed: The Toxic Chemistry of Everyday Products and What's at Stake for American Power* (White River Junction, VT: Chelsea Green Publishing, 2007), 44. The exact mechanisms by which DEHP does damage are not entirely clear, but studies have shown that the chemical can suppress fetal cells that synthesize testosterone; knock out pathways between nurse cells and germ cells that foster the creation of sperm; and reduce production of another growth factor critical in

building the reproductive tract. See K. L. Howdeshell et al., "Mechanisms of Action of Phthalate Esters, Individually and in Combination, to Induce Abnormal Reproductive Development in Male Laboratory Rats," *Environmental Research* 108 (2008): 168–76.

researchers at the Environmental Protection Agency: National Research Council, *Phthalates and Cumulative Risk Assessment,* 2008, 5. Also see L. E. Gray Jr. et al., "Perinatal Exposure to the Phthalates DEH, BBP and DINP, but Not DEP, DMP or DOTP Alters Sexual Differentiation of the Male Rat," *Toxicology Science* 58 (December 2000): 350–65.

98 *female rat pups:* B. J. Davis et al., "Di-(2-Ethylhexyl) Phthalate Suppresses Estradiol and Ovulation in Cycling Rats," *Toxicology and Applied Pharmacology* 128 (1994): 216–23, cited in Wargo et al., "Plastics That May Be Harmful," 40.

even very small amounts: See L. Øie, L. G. Hersoug, and J. O. Madsen, "Residential Exposure to Plasticizers and Its Possible Role in the Pathogenesis of Asthma," *Environmental Health Perspectives* 105 (1997): 972–78, cited in Wargo et al., "Plastics That May Be Harmful," 41–45.

The effects in rats are mirrored: Author interviews with Swan; Joel Tickner, University of Massachusetts at Lowell, Center for Sustainable Production, October 2009; Russ Hauser, physician/researcher at Harvard School for Public Health, October 2009; Rebecca Sutton, senior scientist, Environmental Working Group, September 2009. See also Leonard Paulozzi, "International Trends in Rates of Hypospadias and Cryptorchidism," *Environmental Health Perspectives* 107 (April 1999): 297–302. Danish researchers have argued that such symptoms are connected, part of a condition they called testicular dysgenesis syndrome, which they traced to errors in the development of fetal testes, caused by either genetic defects or environmental factors, such as exposure to hormone-disrupting chemicals. In a paper published in 2001, they contended the syndrome is fairly common, estimating that as many as one in twenty Danish men have at least one or two symptoms. N. E. Skakkebæk et al., "Testicular Dysgenesis Syndrome: An Increasingly Common Developmental Disorder with Environmental Aspects," *Human Reproduction* 16 (May 2001): 972–78.

at least 80 percent of Americans: B. Blount et al., "Levels of Seven Urinary Phthalate Metabolites in a Human Reference Population," *Environmental Health Perspectives* 108 (October 2000): 979–82; M. Silva et al., "Urinary Levels of Seven Phthalate Metabolites in the U.S. Population from the National Health and Nutrition Examination Survey (NHANES) 1999–2000," *Environmental Health Perspectives* 112 (March 2004): 331–38. In fact, some studies suggest that just about every American carries at least one phthalate in his or her system, but until recently the CDC was able to test for only a few of the chemicals' metabolites.

Researchers have detected phthalates: Wargo et al., "Plastics That May Be Harmful," 39; U.S. EPA, "Action Plan on Phthalates," 2009. It's also been shown that the chemicals can cross the placental barrier.

None of us are exposed: It's estimated most of us are taking in 1 to 30 micrograms per kilogram of body weight every day, which is a daily exposure of about 70 to 2,100 micrograms for someone weighing 155 pounds. CERHR, "Monograph," 1.

many of us are taking in more: German researchers found that nearly one-third of the men and women in the study were exceeding the daily-intake limit established by the EPA. In plastic-happy Taiwan, the percentage was 85 percent. The oral threshold set by the EPA in 1986 is 0.02 mg/kg/day, based on the potential for effects to the liver. Wargo et al., "Plastics That May Be Harmful," 40–46.

people with the highest levels: Blount et al., "Levels of Seven"; D. B. Barr et al., "Assessing Human Exposure to Phthalates Using Monoesters and Their Oxidized Metabolites as Biomarkers," *Environmental Health Perspectives* 111 (July 2003): 1148–51.

99 *Splish-Splash Jesus:* Author interview with Swan.

Such findings: CERHR, "Monograph."

the group that seems to be at greatest risk: Interview with Hauser. For more background on NICU babies' exposure, see Ronald Green et al., "Use of Di(2-ethylhexyl) Phthalate–Containing Medical Products and Urinary Levels of Mono(2-ethylhexyl) Phthalate in Neonatal Intensive Care Unit Infants," *Environmental Health Perspectives* 113 (September 2005): 1122–25. CERHR, "Monograph"; Julia Barrett, "NTP Draft Brief on DEHP," *Environmental Health Perspectives* 114 (October 2006): A580–81.

A really sick baby: Luban et al., "I Want to Say," 504.

people of any age undergoing procedures: CERHR, "Monograph," 2.

newborns are underdeveloped: Many scientists consider fetuses, infants, and young children in general to be especially vulnerable to harm from chemicals because their organ systems, metabolic pathways, and hormonal systems are all still developing. Young children also breathe more air and consume more food and drink per pound of body weight, which increases their relative exposure to chemicals in the environment. The National Academy of Sciences in 1993 recognized the special susceptibility of the very young to industrial chemicals in a report on pesticides. Wargo et al., "Plastics That May Be Harmful," 9–10.

100 *We may all be a little plastic:* In an interview with the author, physician/researcher Russ Hauser described the difficulty of drawing practical conclusions from the knowledge that a patient is exposed to phthalates. He has done epidemiological studies showing a correlation between phthalate levels and male infertility. Yet he said the nature of that connection is still too uncertain to be of use in his clinical practice, where he works with infertile couples. While he tests their levels of phthalates and other chemicals used in plastics, he rarely shares the results with them because they are difficult to interpret. It's not like dealing with a known risk such as mercury, he said. If one of his patients had high mercury levels, he could tell him or her how that will affect health and how to avoid mercury in the diet. But with phthalates, he said, "I can't even interpret the level in their urine. If they're forty or eighty or a hundred and twenty parts per billion—does that really impart differences in risks? There's just not enough data [to know.]" Plus,

he added, "We don't want to make someone who's anxious about having a child even more anxious about completely changing their lifestyle."

she measured phthalate levels: Author interview with Swan; Shanna Swan et al., "Decrease in Anogenital Distance Among Male Infants with Prenatal Phthalate Exposure," *Environmental Health Perspectives* 113 (August 2005): 1056–61.

Swan then decided to look at: Author interview with Swan; Swan et al., "Prenatal Phthalate Exposure and Reduced Masculine Play in Boys," *International Journal of Androgyny* 33 (April 2010): 259–69.

102 *Other epidemiological findings:* For overviews of some of those findings, see Wargo et al., "Plastics That May Be Harmful"; John Meeker et al., "Phthalates and Other Additives in Plastics: Human Exposure and Associated Health Outcomes," *Philosophical Transactions of the Royal Society B* 364 (July 2009): 2097–113; Russ Hauser et al., "Phthalates and Human Health," *Occupational Environmental Medicine* 62 (November 2005): 808–18.

girls may also be affected: A study of Puerto Rican girls found high levels of DEHP in more than two-thirds of the girls with premature sexual development and early breast development compared with only about one in five of the subjects with normal puberty. This study has been criticized for possible failure to control laboratory contribution of DEHP to reported tissue concentrations. I. Colón et al., "Identification of Phthalate Esters in the Serum of Young Puerto Rican Girls with Premature Breast Development," *Environmental Health Perspectives* 108 (September 2000): 895–900. Women with endometriosis had higher blood levels of DEHP in studies from Italy and India; see L. Cobellis et al., "High Plasma Concentrations of Di-(2-ethylhexyl)-phthalate in Women with Endometriosis," *Human Reproduction* 18 (July 2003): 1512–15; and B. S. Reddy et al., "Association of Phthalate Esters with Endometriosis in Indian Women," *BJOG* 113 (May 2006): 515–20. Another Italian study also found associations between phthalate levels and uterine fibroids; see S. Luisi et al., "Low Serum Concentrations of Di-(2-ethylhexyl)phthalate in Women with Uterine Fibromatosis," *Gynecological Endocrinology* 22 (February 2006): 92–95.

A 2010 study suggested: Mike Verespej, "Study Says Phthalates May Harm Newborns' Immune Systems," *Plastics News,* July 22, 2010.

German researchers showed: H. von Rettberg et al., "Use of Di(2-Ethylhexyl) Phthalate–Containing Infusion Systems Increases the Risk for Cholestasis," *Pediatrics* 124 (August 2009): 710–16.

studies involving young marmosets: Author interview with Schettler. See C. McKinnell et al., "Effect of Fetal or Neonatal Exposure to Monobutyl Phthalate (MBP) on Testicular Development and Function in the Marmoset," *Human Reproduction* 24 (September 2009): 2244–54. See also CERHR, "Monograph."

Researchers are still debating: The same inconsistencies arise in bisphenol A studies, where the results can be wildly different depending on, for instance, the strain of lab rat that is used. Some are more estrogen-sensitive than others and thus more likely to show response to the chemical. Critics say that's one factor accounting for the huge difference between industry-sponsored studies and those by independent researchers. The animals used in industry-funded studies have been twenty-five thousand to a hundred thousand times less sensitive to estrogen than other species, according to a 2005 study. That same report found that 94 out of 104 studies funded by the government reported significant effects to bisphenol A exposure, while not one of the 11 industry-funded studies found an effect. Frederick vom Saal et al., "Extensive New Literature."

103   *They did a small pilot study:* K. Rais-Bahrami et al., "Follow-up Study of Adolescents Exposed to Di(2-Ethylhexyl) Phthalate (DEHP) as Neonates on Extracorporeal Membrane Oxygenation (ECMO) Support," *Environmental Health Perspectives* 112 (September 2004): 1339–40.

a study by the organization: Environmental Working Group, "Body Burden — The Pollution in Newborns," July 14, 2005. Accessed at http://www.ewg.org/reports/bodyburden2/execsumm.php.

Gray . . . tested mixtures: Author interview with Earl Gray, October 2009.

104   *The American Chemistry Council's position:* Marion Stanley, ACC, quoted in response to FDA public health notification. Also, author interview with Chris Bryant, Phthalate Esters Group, ACC, October 2008. See also CERHR, "Monograph," which contains comments submitted by the ACC in 2005.

the ACC draws on a standing set of criticisms: Wargo, "Pervasive Plastics." See also phthalates-related press releases listed on the ACC's website at http://www.americanchemistry.com/s_acc/sec_newsroom.asp?CID=206&DID=555.

105   *strategy taken straight from the tobacco industry:* David Michaels, "Doubt Is Their Product," *Scientific American* (June 2005): 96.

106   *The EPA . . . recently announced:* EPA, "Phthalate Action Plan," 2009.

The FDA's only action: Food and Drug Administration, "FDA Public Health Notification: PVC Devices Containing the Plasticizer DEHP," July 2002. Accessed at http://www.fda.gov/MedicalDevices/Safety/AlertsandNotices/PublicHealth Notifications/UCM062182.

base their safety assessments: Author interview with Schettler. See also Health Care Without Harm, "Aggregate Exposures to Phthalates in Humans," July 2002. Accessed at http://www.noharm.org/lib/downloads/pvc/Agg_Exposures_to_ Phthalates.pdf.

"a scientifically improbable smoking gun": Schapiro, *Exposed*, 52.

107   *"nearly all chemicals in commerce":* Wargo, "Pervasive Plastics." An expanded version of his analysis can be found in his book *Green Intelligence: Creating Environments That Protect Human Health* (New Haven, CT: Yale University Press, 2009).

at least sixteen thousand: The estimate was made by the EPA and cited in Michael Wilson and Megan Schwarzman, "Toward a New U.S. Chemicals Policy: Rebuilding the Foundation to Advance New Science, Green Chemistry, and Environmental Health," *Environmental Health Perspectives* 117 (August 2009): 1202–9. The 70 percent estimate was cited by Joel Tickner in an e-mail to the author, October 2010.

European regulators "act on the principle of preventing": Schapiro, *Exposed,* 52.

Europeans began limiting DEHP: In 2001, the EU classified DEHP as "toxic" and also barred its use in cosmetics and in all children's products. Starting in 1998, American toy manufacturers voluntarily removed DEHP from teethers, rattlers, and other toys that could be mouthed by children under the age of three. Yet that doesn't prevent importation of toys containing phthalates, which is significant, since 80 percent of toys sold in the United States are produced in and imported from China, where there's no restriction on the use of DEHP.

The agency charged: Health Care Without Harm, "The Weight of the Evidence on DEHP." Accessed at http://www.noharm.org/lib/downloads/pvc/Weight_of_Evidence_DEHP.pdf.

108 *approach taken by Massachusetts:* Author interviews with Liz Harriman, Pam Eliason, and Greg Morose of the Toxics Use Reduction Institute, the organization established to educate companies about nontoxic alternatives and help them comply with the state law, October 2009.

   national "plastics control law": Wargo, "Pervasive Plastics."

109 *Safreed was a nurse:* Author interview with Paula Safreed, former nurse in NICU at Brigham and Women's Hospital, September 2009. Also, author interview with Julianne Mazzawi, assistant manager of the neonatal intensive care unit at Brigham and Women's Hospital, September 2009.

   Discovery of that fact: Gary Cohen, Health Care Without Harm, Skoll video, accessed online at: http://www.youtube.com/watch?v=tR4Pz9qwRyo. Information on the group's history also gathered through author interviews with Tickner; Mark Rossi, research director, Center for Clean Production, September 2009; and Stacey Malkin, former spokeswoman for Health Care Without Harm, September 2007.

110 *about 120 of the more than 5,000 hospitals:* Health Care Without Harm, "List of Hospitals Undertaking Efforts to Reduce PVC or DEHP." Accessed at http://www.noharm.org/lib/downloads/pvc/List_of_Hosps_Reducing_PVC_DEHP.pdf. Most are located in California, the Pacific Northwest, and New England. See also Laura Landro, "Hospitals Go 'Green' to Cut Toxins, Improve the Patient Environment," *Wall Street Journal,* October 4, 2006.

   what's really driven change: Author interviews with Rossi, Tickner.

111 *Hexamoll DINCH:* Author interview with Patrick Harmon, September 2009.

   they constitute only about a quarter: Estimate by Mark Ostler, manager of materials technical services at Hospira, in interview with author, December 2009.

112 *there are also applications for which no alternatives:* That fact recently put Luban and Short in the strange position of having to lobby against a local proposal to outlaw the use of medical devices containing DEHP. Even though both doctors are concerned about the health impacts of the chemical, had the law passed, said Luban, "no one [in DC] would have been able to receive a blood transfusion."

113 *recent study of the vinyl-free Brigham NICU:* A. M. Calafat et al., "Exposure to Bisphenol A and Other Phenols in Neonatal Intensive Care Unit Premature In-

fants," *Environmental Health Perspectives* 117 (April 2009): 639–44; author interview with Steve Ringer, chief of newborn medicine at Brigham and Women's Hospital, August 2009.

5. Matter Out of Place

115 *I was talking to a researcher:* Author interview with David Karl, professor of oceanography, University of Hawaii, July 2009.

116 *hundreds of beached lighters:* Author interview with Seba Sheavly, marine-debris consultant who has done studies of debris on Midway beaches, January 2010.

"could stock the check-out counter": Charles Moore, "Trashed," *Natural History* 112 (November 2003).

He routinely finds: Author interview with John Klavitter, U.S. Fish and Wildlife Service, April 2010.

117 *One dead chick:* Curtis Ebbesmeyer and Eric Scigliano, *Flotsametrics and the Floating World: How One Man's Obsession with Runaway Sneakers and Rubber Ducks Revolutionized Ocean Science* (New York: HarperCollins, 2009), 212.

major threat is plastic: Heidi Auman et al., "Plastic Ingestion by Laysan Albatross Chicks on Sand Island, Midway Atoll, in 1994 and 1995," in G. Robinson and R. Gales, eds., *Albatross Biology and Conservation* (Chipping Norton, Australia: Surrey Beatty and Sons, 1997), 239–44.

In one two-month cleanup: Webpage of Midway volunteers, accessed at http://kms.kapalama.ksbe.edu/projects/2003/albatross/.

a new factor likely contributing: Author interview with Klavitter; Auman, "Plastic Ingestion"; also, Kenneth R. Weiss, "Plague of Plastic Chokes the Seas," *Los Angeles Times,* August 2, 2006.

118 *more than 260 species of animals:* Jose G. B. Derraik, "The Pollution of the Marine Environment by Plastic Debris: A Review," *Marine Pollution Bulletin* 44 (September 2002): 842–52. See also David Laist, "Impacts of Marine Debris: Entanglement of Marine Life in Marine Debris Including a Comprehensive List of Species with Entanglement and Ingestion Records," in J. M. Coe and D. B. Rogers, eds., *Marine Debris — Sources, Impacts, and Solutions* (New York: Springer Verlag, 1997), 99–139.

"One of the most ubiquitous": David Barnes et al., "Accumulation and Fragmentation of Plastic Debris in Global Environments," *Philosophical Transactions of the Royal Society B* 364 (July 2009): 1987.

119 *a hazard for wildlife:* A recent news story described how German hedgehogs, having developed a taste for McDonald's McFlurrys, kept getting their heads caught in the dessert containers' convex lids and then starving to death. To its credit, McDonald's changed the lid design to one that was more hedgehog-friendly. Patrick MacGroarty, "McDonald's Redesigns Deadly Lids," *Der Spiegel,* February 27, 2008. Accessed at http://www.spiegel.de/international/zeitgeist/0,1518,538125,00.html.

Even plastics that are properly thrown away: Emma L. Teuten et al., "Transport and Release of Chemicals from Plastics to the Environment and Wildlife," *Philosophical Transactions of the Royal Society B* 364 (July 27, 2009): 2027–45.

No one knows for sure: The figure of 13,000 pieces per square kilometer is cited in a report issued by the UN Environment Program, "Marine Litter: An Analytical Overview," 2005, 4. Accessed at www.unep.org/regionalseas/publications/Marine_Litter.pdf. The source of the 3.5 million pieces per square kilometer is Richard Thompson et al., "Plastics, the Environment and Human Health: Current Consensus and Future Trends," *Philosophical Transactions of the Royal Society B* 364 (July 27, 2009): 2155.

One expert estimated: Andrady, "Applications and Societal Benefits," 1981. Andrady calculated that 0.2 to 0.3 percent of all plastics produced worldwide each year end up in the ocean, which would be 1.0 to 1.6 billion pounds. Cod fishery information from UN Food and Agricultural Organization, *Fishery and Aquaculture Statistics*, 2007 (Rome, 2009), 12. Accessed at http://www.fao.org/fishery/publications/yearbooks/en.

Researchers first began noting: The first accounts came from reports of plastic fragments found in carcasses of sea birds collected from shorelines. Thompson, "Plastics, the Environment," 2154.

the volume of plastic fibers: Richard Thompson, "Lost at Sea: Where Is All the Plastic?" *Science* 304 (May 2004): 838; also Ebbesmeyer and Scigliano, *Flotsametrics*, 203.

may be leveling off: Thompson, "Plastics, the Environment," 2155; author interviews with Seba Sheavly, marine-debris consultant, January and June 2010; author interview with Kara Lavender Law, chief scientist Sea Education Association, an educational nonprofit based in Woods Hole, Massachusetts, August 2009. According to Law, since the 1980s the group has run research vessels on the same route through the North Atlantic and the Caribbean, sampling plankton with special nets that have 0.3-millimeter mesh. The trawls have always pulled up plastic. There's no sign the amounts are increasing, but in the 1980s the nets collected large number of preproduction pellets, and today the trawls are more likely to bring in fragments of postconsumer plastics.

120   *Plastics weren't something people could make or fix:* Susan Strasser, *Waste and Want: A Social History of Trash* (New York: Metropolitan Books, 1999), 267.

"rosily astronomical": "Plastics in Disposables and Expendables," *Modern Plastics* (April 1957): 94.

121   *as a speaker at a 1956 conference:* Ibid., 93.

such products were a tough sell: Heather Rogers, *Gone Tomorrow: The Hidden Life of Garbage* (New York: New Press, 2005), 109.

when vending machines: "Plastics in Disposables," *Modern Plastics* (April 1957): 96.

Life magazine celebrated: "Throwaway Living," *Life,* August 1, 1955.

half of all plastics: Jefferson Hopewell et al., "Plastics Recycling: Challenges and

Opportunities," *Philosophical Transactions of the Royal Society B* 364 (July 2009): 2115.

122 *The lighter was the brainchild:* Gordon McKibben, *The Cutting Edge: Gillette's Journey to Global Leadership* (Boston: Harvard Business Press, 1998), 101–4.

"the attractive toss-away": Lee Daniels, "Gillette Gives Up on Cricket," *New York Times,* October 5, 1984.

the two companies duked it out: McKibben, *Cutting Edge.* The lighter was actually just one product in a long and legendarily nasty war over consumer disposables waged between the two companies. They first battled over ballpoints, with Bic eventually beating out Gillette's Paper Mate pen. Then came the lighter fight, which Bic quickly won; by 1984, when Gillette sold off Cricket to a Swedish company, Bic controlled 55 percent of the estimated $325 million market (Daniels, "Gillette Gives Up"). The companies also fought over disposable razors, which each introduced in the mid-1970s. "We beat them in ballpoints, we beat them in lighters, and there's no reason we can't beat them here," one advertising executive told the *New York Times* when Bic launched its throwaway shaver in 1976 (Philip Dougherty, "Bic Pen Challenges Gillette on Razors," *New York Times,* October 29, 1976). But in that battle Gillette has retained the edge.

worldwide annual sales: McKibben, *Cutting Edge,* 102. Yet as rich as the market was, neither Bic nor Gillette made big profits because, in the aggressive battle for dominance, each company kept dropping prices to undercut the other.

123 *smoking is on the rise:* Judith McKay and Michael Eriksen, *The Tobacco Atlas,* The World Health Organization, 2002, 91. Accessed at http://www.who.int/tobacco/en/atlas38.pdf.

Bic now has markets: Information accessed from Bic website, www.Bicworld.com.

In exports alone, China sold: Allen Liao, "Born of Fire: China's Lighter Manufacturing Industry," *Tobacco Asia,* Q2, 2009. Accessed at http://www.tobaccoasia.net/previous-issues/features/39-featured-articles-q2-2009/128-born-of-fire-china-is-lighter-manufacturing-industry. The business still has high enough volumes that one Chinese manufacturer of disposable lighters advertises it will not take any orders of fewer than 500,000 units.

"a bridge between metal": DuPont Delrin website, accessed online at www2.dupont.com/Plastics/en_US/Products?delrin/Delrin.html.

125 *"of little practical consequence":* Anthony Andrady, "Plastics and Their Impact in the Marine Environment," Proceedings of the International Marine Conference on Derelict Fishing Gear and the Ocean Environment, August 2000. According to Andrady, how long it takes for a polymer to biodegrade also depends on the immediate environment — whether it is wet or dry, cold or warm, exposed to a lot of sun or not — and on the type of polymer. In one study, researchers incubated polyethylene in a culture of live bacteria. After a year, less than 1 percent was gone. (Author interview with Andrady, October 2009, and described in Weisman, *The World Without Us,* 127.)

In the ocean, that process slows: The one exception is expanded polystyrene, a.k.a. Styrofoam, which is more persistent on land than in the ocean, according to Andrady. On land, it reacts with UV radiation to form a yellowish surface layer that helps hold the material together. In water, the film washes off, and the material quickly disintegrates into microscopic pieces. The material is no longer visible, but that doesn't count as degradation.

"the sea floor": Murray R. Gregory, "Environmental Implications of Plastic Debris in Marine Settings—Entanglement, Ingestion, Smothering, Hangers-on, Hitchhiking and Alien Invasions," *Philosophical Transactions of the Royal Society B* 364 (July 27, 2009): 2017.

"assembly of ghosts": Ibid.

126 *I'd been told about the beach:* Author interviews with Judith Selby Lang and Richard Lang, March and November 2008.

converging currents throw up: Ebbesmeyer and Scigliano, *Flotsametrics,* 200–201.

127 *when fishing fleets began switching:* Gregory, "Environmental Implications," 2014.

Plastic makes up only about 10 percent: Barnes et al., "Accumulation and Fragmentation."

128 *beach surveys around the world:* Derraik, "Pollution of the Marine Environment." In some places it's even higher. One survey of Cape Cod beaches and harbors found 90 percent of the debris was plastic.

"lubricant of globalization": Author interview with Charles Moore, May 2009.

what's also striking is the uniformity: Ocean Conservancy, *A Rising Tide of Ocean Debris and What We Can Do About It: 2009 Report.* There's also a lot of nonplastic debris collected, including paper bags, glass bottles, tin cans, and pull-tabs.

cigarette butts: Ebbesmeyer quoted in Hohn, "Moby-Duck."

Disposable lighters aren't far behind: Ocean Conservancy, *Rising Tide.* The United States reported finding 18,555.

ten-million-square-mile: Hohn, "Moby-Duck." For a good description of the gyre, see Ebbesmeyer and Scigliano, *Flotsametrics.*

129 *"As I gazed from the deck":* Moore, "Trashed."

cargo containers lost at sea: It's estimated two thousand containers go overboard annually, down from ten thousand. Ebbesmeyer and Scigliano, *Flotsametrics,* 205.

an area that he estimated: Moore, "Trashed"; author interview with Moore.

130 *a host of misperceptions:* One misconception has to do with a study by Moore in which he reported finding that the mass of plastic he'd pulled out of the vortex outweighed the mass of plankton there six to one. The statistic is often repeated, and yet experts point out that it's misleading. The vortex is an ocean desert—the particular features that form it deter an abundance of marine life, which means one wouldn't expect to find much plankton in the area. Moreover, plankton consist mostly of water; once dried out—as was done in Moore's study—their mass seems smaller than it actually is.

an "eighth continent": Editorial, "Our Plastic Legacy Afloat," *New York Times,* August 26, 2009.

as voyagers there have discovered: These descriptions come from interviews with members of Project Kaisei, an expedition that journeyed to the gyre in the summer of 2009 to document the extent of debris and investigate possibility of cleaning it up. Author interviews with Miriam Goldstein, Andrea Neal, Dennis Rogers, Nelson Smith, and Doug Woodring, August 2009.

vortex "is not a static environment": Author interview with Sheavly.

131 *more harm than good:* The warning was made by Andrea Neal, who eventually overcame her reservations about the project and joined as one of its scientific advisers. Author interview with Neal, July 2009; author interview with Woodring, August 2009.

There are at least five: The number is subject to debate. Based on his flotsam studies and computer modeling, Ebbesmeyer calculates there are eleven distinct ocean gyres and eight high-pressure vortexes. Researchers Peter Niiler and Nikolai Maximenko tracked the paths of fifteen thousand research floats to trace ocean currents around the globe, and their resulting map shows just five areas where currents converge in a vortex. "Tracking Ocean Debris," *IPCR Climate* 8 (2008): 14–16; author interview with Peter Niiler, Scripps Institution of Oceanography, July 2009.

One gyre can be found in the North Atlantic: Author interview with Law. The 2010 research trip to the North Atlantic gyre by her organization, SEA, was the first backed by federal funding. Results of the trip were reported in Kara Lavender Law et al., "Plastic Accumulation in the North Atlantic Subtropical Gyre," *Science* 329 (2010): 1185. See also Melissa Lang, "Fishing for Plastic in the Atlantic," *Boston Globe,* July 14, 2010.

where Ahab's crew: Author interview with Niiler.

132 *"in impaired movement and feeding":* Thompson, "Plastics, the Environment," 2155.

How many are killed a year?: Author interview with Holly Bamford, director of the marine-debris program at the National Oceanographic and Atmospheric Administration, July 2009. The figure seems to have originated from a paper by C. Fowler, "Status of Northern Fur Seals on the Pribilof Islands," submitted to the twenty-sixth annual meeting of the Standing Scientific Committee of the North Pacific Fur Seal Commission, in 2008. It stated: "Debris entanglement is estimated to cause 50,000 to 90,000 deaths per year in the northern fur seal. At least 50,000 deaths are thought to be due to entanglement; the other 40,000 deaths possible entanglement or possibly some unknown factor such as disease." A 2008 article in the London *Times* described how a later misquote of the report, which never mentioned plastic bags, helped fuel anti-plastic-bag feelings. Alexi Mostrous, "Series of Blunders Turned the Plastic Bag into Global Villain," London *Times,* March 8, 2008. Accessed online at http://ad.uk.doubleclick.net/adj/news.timesonline.co.uk/environment;pos=sponsor;sz=143x50;tile=2;yahoo=No;ord=1274124758122?.

no documented basis: Author interview with Bamford. See also NOAA marine-debris website Frequently Asked Questions, accessed at http://marinedebris. noaa.gov/info/faqs.html#6.

cause of injury or death: Derraik, "Pollution of the Marine Environment"; Study on southern fur seals cited in Charles Moore, "Synthetic Polymers in the Marine Environment: A Rapidly Increasing Threat," *Environmental Research* 108 (October 2008); on fulmars, see M. Malloy, "Marine Plastic Debris in Northern Fulmars from the Canadian High Arctic," *Marine Pollution Bulletin* 58 (August 2008): 1501–4.

133 *the Peruvian beaked whale:* "Exfoliating Scrubs Join List of Plastics Harming Whales," *Scotland Sunday Herald,* March 9, 2008.

Leatherback turtles: "Leatherback Turtle Threatened by Plastic Garbage in the Ocean," *Science Daily,* March 16, 2009. Reporting on study by N. Mrosovsky et al., "Leatherback Turtles: The Menace of Plastic," *Marine Pollution Bulletin* 58 (2009): 287.

Endangered humpback whales: Gregory, "Environmental Implications."

British biologist David Barnes: Barnes et al., "Accumulation and Fragmentation." Author interview with Barnes, June 2009.

"Plastic is not just an aesthetic": Barnes, quoted in Thomas Hayden, "Trashing the Oceans," *U.S. News and World Report,* November 14, 2002.

134 *The first conference:* The proceedings of the conference provide a good overview of the concerns about and knowledge of microdebris. See National Oceanic and Atmospheric Administration, *Proceedings of the International Workshop on the Occurrence, Effects and Fate of Microplastic Marine Debris,* September 9 to September 11, 2008, published January 2009. See also Murray Gregory and Anthony Andrady, "Plastics in the Marine Environment," in Andrady, ed., *Plastics and the Environment,* 381–82, and Richard C. Thompson et al., "Our Plastic Age," 1975.

135 *increase in pellets:* Moore, "Synthetic Polymers." In 1991 the Society of Plastics Industry created a program, Operation Clean Sweep, designed to prevent pellet losses. The measures recommended through Clean Sweep are effective at reducing by some 50 percent the amounts that get into the environment. Unfortunately it remains voluntary, and according to Moore, only a small percentage of companies participate.

teensy plastic beads as scrubbers: Author interview with Richard Thompson, University of Plymouth, September 2009.

In one series of lab experiments: Author interview with Thompson; Weisman, *World Without Us,* 116.

similar feeding study with mussels: Described in Thompson, "Plastics, the Environment," 2156.

In a 2008 return trip to the Pacific gyre: The results were described on the website of Moore's Algalita Marine Research Foundation in "Update on Fish Ingestion Study," September 2009, at http://www.algalita.org/bispap-ingestion-update-9-09.html. Also in David Ferris, "Message in a Bottle," *Sierra* magazine (May/June 2009).

136 *Japanese researchers:* Teuten et al., "Transport and Release of Chemicals," 2035–37.
that's no surprise to scientists: Author interview with Bamford.
Indeed, Takada argues: Author interview with Hideshige Takada, February 2009.
See also Yuko Ogata et al., "International Pellet Watch: Global Monitoring of Persistent Organic Pollutants (POPs) in Coastal Waters. Initial Phase Data on PCBs, DDTs and HCHs," *Marine Pollution Bulletin* 58 (October 2009): 1437–46.
I packed up a hundred and fifty pellets: All but thirty-three of the pellets were polyethylene (the rest were polypropylene), and those were the ones he analyzed. Author e-mail correspondence with Takada, January 2010.

137 *More than 180 species:* Emma L. Teuten, "Microplastic–Pollutant Interactions and Their Implication in Contaminant Transport to Organisms," presented at the International Workshop on the Occurrence, Effects and Fate of Microplastic Marine Debris, September 2008.
Thompson placed lugworms: Author interview with Thompson. The studies are described in Teuten et al., "Transport and Release of Chemicals," 2038.
Teuten fed sea birds pellets: Ibid., 2040.
Hans Laufer found alkyphenols: Author interview with Hans Laufer, August 2008. Also see Abbie Mitchell, "Plastic in the Ocean Hurts Lobster," *Nova News Now,* July 3, 2008, accessed at http://www.novanewsnow.com/article-227641-Plastic-in-the-ocean-hurts-lobster.html.

138 *the Bradford, Pennsylvania, company:* Author e-mail correspondence with Pat Grandy, communications manager, Zippo Corporation, September 2009.
collectors have little interest: Author interviews with Judith Sanders, On the Lighter Side; Ted Ballard, National Lighter Museum, June 2009.

6. Battle of the Bag

141 *"there is simply no justification":* Achim Steiner, UN undersecretary-general and UNEP executive director, press release for issuance of UN report on marine debris, June 10, 2009; http://www.unep.org/Documents.Multilingual/Default.asp?DocumentID=589&ArticleID=6214&L=EN&T=LONG.
more than two hundred anti-bag measures: Mike Verespej, "Plastic Bag Industry in Fight of Its Life," *Plastics News,* March 16, 2009.
an emblem "of waste and excess": Belinda Luscombe, "The Patron Saint of Plastic Bags," *Time,* July 27, 2008.
"the plastics industry has helped turn us into": Stephanie Barger, executive director of the Costa Mesa–based Earth Resource Foundation and founder of the Campaign Against the Plastic Plague, quoted in Steve Toloken, "Plastics' Image Problem," *Plastics News,* August 6, 2007.

142 *"Not a single solid market":* "No Easy Years Ahead," *Modern Plastics* (June 1956): 5.
About half of all goods: Mike Verespej, "Even Going Green Requires Cutting Costs," *Plastics News,* December 22, 2008.
one of every three pounds: American Chemistry Council, *Resin Review 2008,* 51.

"bag panic": Meikle, *American Plastic,* 249–50.

the head of the Society of the Plastics Industry: Mildred Murphy, "Plastic Industry to Warn on Bags," *New York Times,* June 18, 1959.

143 *As Jerome Heckman:* Jerome Heckman, "Heckman Shares Plastics Past," *Plastics News,* April 17, 2000.

Mobil began eyeing: The account of Mobil's development of T-shirt bags comes from author interview with Bill Seanor, now with Overwraps Packaging, Inc., September 2008.

the classic brown paper bag: According to Diana Twede, a packaging scientist and historian of the field, the paper shopping bag appeared in the mid-nineteenth century when an enterprising grocer in Bristol, England, figured out he could glue together some paper to make a bag and gain a valuable twofer: a convenient way for his customers to carry their goods home, and a free space to advertise his store. Until the process became automated, after the Civil War, shopkeepers passed their off-hours sitting in the backs of their stores gluing paper together to make bags. Still, it wasn't until the rise of car culture, suburbs, and self-serve grocery stores in the decades after World War II that the brown paper bag became a staple at the checkout stand. Author interview with Diana Twede, Michigan State University at Lansing, August 2008.

mainly in Europe: Europe was more open to plastic bags than the United States, which, owing to the abundance of timber, has always enjoyed lower paper prices.

Sten Thulin had come up with a design: "Design Landmarks," *European Plastics News,* October 1, 2008. Author interview with Chris Smith, editor of *European Plastics News,* September 2008. Inventors had long been trying to do the same, but Thulin was the first to develop a workable method of folding, welding, and die-cutting a flat tube of plastic. Or one of the first — there was a Finnish patent issued around the same time.

company's early foray into the production: Mobil brought a small factory in Florence, Italy, that was producing T-shirt bags and sent Seanor there to rev up production for the American market.

144 *shoppers were underwhelmed:* Another problem was the polymer Mobil used for its bags: a linear-low-density polyethylene that easily stretched and tore. After a few years, bag manufacturers switched to a hardier plastic, high-density polyethylene (HDPE), that was better suited to carting heavy loads. The switch to HDPE "cut the legs out from under" Mobil's bag business, said Seanor. The company was too invested in LLDPE to make the change, and eventually it ceded the T-shirt-bag market to a crop of new companies that were using HDPE to make bags.

"Check out the Sack": Flexible Packaging Association, "An Industry Takes Shape: The FPA Story, 1950–2000," 28.

most persuasive factor: The paper industry did not give up without a fight. It defended paper bags on environmental grounds. With environmental concerns focused on threats of overflowing landfills, paper did have a better story to tell: it was a biodegradable material and more readily recycled. For many consumers, that was an important consideration, and the American Paper Institute was more than happy to exploit their concerns. The group rallied the nine-thousand-chapter-strong General Federation of Women's Clubs to fight for what one of the group's leaders called "the all-American paper bag." "We have to speak out against over-packaging and we must demand environmentally sound package choices," GFWC leader Margaret England declared. George Makrauer, "Plastic Bag Wars and Politics," unpublished history shared with author.

"Once we started getting the Krogers": Author interview with Peter Grande, September 2008.

people used somewhere between: That's the estimate used by Reusablebags.com.

145 *The average American:* Based on estimates that Americans consume about 90 billion bags a year and the U.S. population is about 300 million.

as DuPont had encouraged: Meikle, *American Plastic,* 250.

"clear plastic bags": "Cleaning Plastic Bags," *New York Times,* May 8, 1956. When Imperial Chemical Company introduced the first polyethylene bags, the bags were costly enough that each pack came with instructions as to how they could be washed for reuse. Author interview with Chris Smith.

the average American throws out: Daniel Imhoff, *Paper or Plastic: Searching for Solutions to an Overpackaged World* (San Francisco: Sierra Club Books, 2005), 9; EPA, *Municipal Solid Waste Generation, Recycling and Disposal in the United States: Facts and Figures for 2008.* It is perhaps not a coincidence that between 1960 and 1980 the amount of solid waste in the United States tripled.

146 *"It provided battles":* Daniel Weintraub, "Into the Battle," *Orange County Register,* June 11, 2000.

According to Murray: Author interviews with Mark Murray, September and November 2008, June 2010.

147 *take up much less space:* In studies of various landfills by William Rathje's Garbage Project, plastic packaging and goods consistently took up 20 to 24 percent of all garbage when first sorted, and 16 percent after it had been compacted. William Rathje and Cullen Murphy, "Five Major Myths About Garbage and Why They're Wrong," *Smithsonian* (July 2002). More recent waste studies by the EPA show plastics constitute about 12 percent of the total waste stream, while paper makes up 31 percent.

bag ban proposed in Fairfield: Kirk Lang, "Fairfield Panel Recommends Trashing Plastic Bags," www.connpost.com, August 13, 2009.

Archaeologist William Rathje: Rathje's work in landfills supplied valuable ammunition for the plastics industry in the 1980s. The industry sponsored his research and widely disseminated his findings. Plastic-bag makers used his findings to bolster their case against paper bags, said Seanor. "We probably gave his videos and stuff like that to every retailer in the country."

The environmental implications: For instance, a 2010 proposal to ban bags in California was sponsored by Santa Monica's Heal the Bay, and environmental groups were a major part of the coalition that supported the measure.

148 *Zero waste:* Author interview with Robert Haley, Zero Waste manager, San Francisco Department of Environment, October and November 2008. Also Elizabeth Royte, *Garbage Land: On the Secret Trail of Trash* (Boston: Little, Brown, 2005), 255.

at least four times: Robert Lilienfeld, "Revised Analysis of Life Cycle Assessment Relating to Plastic Bags," *ULS Report,* March 2008, accessed at http://use-less-stuff.com/research.htm.

149 *forty-six-billion-dollar tourist industry:* Green Cities California, *Master Environmental Assessment on Single-Use and Reusable Bags,* issued March 2010, 15.

explained Leslie Tamminen: Author interview with Leslie Tamminen, July 2010.

2008 international beach cleanup: Ocean Conservancy, *A Rising Tide,* 9. All told, bags made up 12 percent of all the litter collected, second only to cigarette butts. Interestingly, one of the sponsors of these annual beach cleanups is Dow Chemical, a leading producer of polyethylene resin, the material from which T-shirt bags are made.

A study in Los Angeles: Los Angeles County Department of Public Works, "An Overview of Carryout Bags in Los Angeles County" (August 2007): 24.

Complying with the mandate: Californians Against Waste, "The Problem of Plastic Bags," CAW website, accessed at http://www.cawrecycles.org/issues/plastic_campaign/plastic_bags/problem.

150 *they had very practical reasons:* Author interviews with Haley.

151 *levied a fifteen-cent fee:* Bags dropped from about 45 percent of all litter to 0.22 percent in 2004, according to Frank Convery et al., "The Most Popular Tax in Europe? Lessons from the Irish Plastic Bags Levy," *Environmental and Resource Economics* 38 (September 2007): 7. South Africa and Hong Kong have adopted similar fees.

about as socially acceptable: Elisabeth Rosenthal, "Motivated by a Tax, Irish Spurn Plastic Bags," *New York Times,* February 2, 2008.

The plastax generated: Convery et al., "Most Popular Tax," 9. In later years, the fee was increased.

"it would be politically damaging": Ibid., 2. Industry often points out that in the wake of the plastax, Irish imports of plastic garbage bags and polyethylene film increased substantially — from 26.3 million metric tons in 2002 to 31.6 million in 2006. Bag defenders often portray that increase as an "unintended consequence" of the ban. But that sidesteps the fact that the point of the tax wasn't to eliminate all plastic bags, just the ones contributing to litter. And garbage bags don't have the litter impact of the lightweight shopping bags.

152 *the plastics industry:* Tim Shestek, lobbyist for the American Plastics Council, quoted in Suzanne Herel, "Paper or Plastic: Pay Up," *San Francisco Chronicle,* November 20, 2004.

end run around the city: This occurred during a period in which the grocers had agreed to voluntarily reduce their distribution of plastic bags if Mirkarimi would

table the proposed fee for a year. The understanding was that if the stores succeeded in significantly decreasing the numbers of bags they gave out, the city
would shelve its fee plan. Author interviews with Haley, Ross Mirkarimi, San
Francisco Board of Supervisors, December 2008. See also Charles Goodyear,
"Deal to Reduce Plastic Bag Use Hits the Skids," *San Francisco Chronicle,* January
18, 2007.

the explosive effect of limiting cities' choices: Author interviews with Murray, Haley,
and Mirkarimi.

153 *cities around the country:* See, for instance, Mark McDonald, "Dump the Plastic
Bag?" *Philadelphia Daily News,* September 18, 2007; Carolyn Shapiro, "The Flap
Over Plastic Shopping Bags," *The Virginian-Pilot,* March 10, 2008; and Kyle Hopkins, "Tundra Trash: Bethel Prohibits Plastic Bags," *Anchorage Daily News,* July
21, 2009.

154 *Walmart became the new FDA:* Marc Gunther, "Wal-Mart: The New FDA," *Fortune,* July 16, 2008.

Sixty Minutes phenomenon: Roger Bernstein, vice president of state affairs and
grass roots, American Chemistry Council, interview with author, October 2008.

"righteous simplicity": Joe Eskenazi, "Baggage," *San Francisco Weekly,* January 5,
2009.

"This is not a wacko group": "Purge the Plastic Plague," *Die-Line: The Newsletter of
the California Film Extruders and Converters Association,* January 2005.

a spokesman for the group commandeered: Steve Toloken, "APC, Calif. Clash on
Marine Debris Issue," *Plastics News,* December 6, 2004.

industry polls showing: As late as 2007, an industry poll found that half the re-
spondents considered the bag fights and other battles over plastic packaging that
were occurring in California as irrelevant outside the state or an overreaction.
The fact that so many in the business "think we don't have a serious problem
. . . makes me wonder if the real problem is literacy, complacency or both," Pete
Grande, a Los Angeles–based maker of plastic bags, complained in an editorial
in the newsletter *Die-Line* in March 2007. Most of the California bag makers who
were speaking out about marine debris did not actually make T-shirt bags; they
made bags for produce or restaurants or department stores. But they feared any
backlash against T-shirt bags would eventually reach their products.

155 *"Our industry has been really slow":* Author interview with Robert Bateman, September 2007.

annual revenues of more than $120 million: According to its 2006 tax return, rev-
enues were $122.8 million. And in 2008, the new head of the ACC referred to its
budget being four times that of his prior employer's $30 million operating budget.
Mike Verespej, "New ACC Head Discusses Challenges Ahead," *Plastics News,* July
28, 2008.

Society of the Plastics Industry: In 2007, the SPI's operating budget was $7 million,
down from $30 million a decade before; the group has since made further staffing

and budget cuts. See Mike Verespej, "SPI Shake-Up Concerning Some Members," *Plastics News,* August 13, 2007, and "SPI Cuts Staff, Reorganizes Amid Downturn," *Plastics News,* July 13, 2009.

156 *"their ox wasn't getting gored":* Seanor quoted in Mike Verespej, "Plastic Bag in Fight for Its Life," *Plastics News,* March 16, 2009.

about $1.2 billion: Estimate by Isaac Bazbaz, head of Superbag Corp., one of the nation's largest T-shirt-bag manufacturers. Author interview with Bazbaz, August 2008.

they called a meeting: This account of the meeting is based on author interviews with Seanor and Bazbaz. The bag operations of Vanguard and Sunoco were subsequently bought by Hilex Poly, which is based in South Carolina.

various of the companies contributed more: Isaac Bazbaz of Superbag reportedly spent more than one million dollars on PBA efforts. Brett Clanton, "Bag Makers Defend Plastic," *Houston Chronicle,* December 1, 2007.

a full-time job for Johnson: It was also an eye-opening experience, according to Seanor and other colleagues. He'd begun the job hostile to the environmentalists' complaints about plastic pollution. After a year, he'd begun thinking that the plastics industry as a whole should pay into a fund devoted to mitigating the problem of plastic debris in the ocean. He died before he had a chance to develop the idea.

157 *plastics-related legislation:* Keith O'Brien, "In Praise of Plastic," *Boston Globe* magazine, September 28, 2008.

"We are at the tipping point": Quoted in Meg Kissinger and Susanne Rust, "Plastics Fights Back with PR," *Milwaukee Journal Sentinel,* August 22, 2009.

Consumption of paper bags shot up: Eskenazi, "Baggage."

158 *prior success as a campaigner:* Author interviews with Joseph, July 2008 and June 2009. See also Luscombe, "Patron Saint."

a man who knew how to champion: Author interview with Pete Grande.

news article in the London Times: Mostrous, "Series of Blunders."

Life-cycle analyses: At least a half dozen LCAs of bags have been done, in Europe, Australia, South Africa, and the United States. For a roundup and links to the most credible ones, see Lilienfeld, "Revised Analysis," March 2008.

159 *"the only thing civilized":* Tom Robbins, quoted in Diana Twede and Susan E. M. Selke, *Cartons, Crates and Corrugated Board: Handbook of Paper and Wood Packaging Technology* (Lancaster, PA: DEStech Publications, 2005).

160 *"karma's a bitch":* Gene Maddaus, "The Plastic Bag Lives on in Manhattan Beach," *LA Weekly,* January 27, 2010.

such lawsuits, or threats of lawsuits: For instance, the city of Fairfax, a small town of seven thousand people and two grocery stores, was sued when it passed a bag ban in 2007. Town leaders withdrew the proposal rather than endure a court battle or do a full-blown environmental report. But Fairfax leaders found a different way to banish the bags, by putting the ban proposal to voters through a ballot initiative, which doesn't require the same environmental-impact review. Voters approved it in 2008.

161 *it confirmed what Joseph had been saying:* Green Cities California environmental-impact report, 2010.

That finding was surprising: Author interview with Carol Misseldine, May 2010.

I met with him: Author interviews with Roger Bernstein; Keith Christman, senior director of packaging, plastics division, ACC; and Jennifer Killinger, senior director of sustainability and public outreach, ACC, in October 2008.

162 *it took up the fight:* Author interview with Bazbaz. Under the ACC's aegis, the bag makers' group formerly known as the Progressive Bag Alliance got a new name: the Progressive Bag Affiliates. Strangely, the SPI has mostly stayed out of the battle.

163 *Plastics Make It Possible:* Ronald Yocum and Susan Moore, "Challenge 2000: Making Plastic a Preferred Material," in Rosato et al., *Concise Encyclopedia*, 15. The campaign was sponsored by the American Plastics Council, and industry-sponsored polls showed plastics' favorability ratings rose from around 52 percent to the mid-60s range. See also Steve Toloken, "The Plastics Wars," *Plastics News*, March 8, 1999.

Aggressive industry lobbying: For instance, the industry's tough tactics succeeded in preventing Suffolk County, New York, from implementing the nation's first ban on plastic packaging, including plastic bags. After the law was passed in 1988, manufacturers sued, tying it up in litigation for four years. The county won in court, but by then recycling programs had been put in place, the public had lost interest, and the county decided to drop the ban. John McQuiston, "Suffolk Legislators Drop a Ban on Plastic Packaging for Foods," *New York Times*, March 9, 1994.

major public relations effort: Information about the fashion shows and other programs can be found at http://www.plasticsmakeitpossible.com. Meanwhile the SPI has pledged to launch a major ten-million-dollar Internet campaign, Imagine the Possibilities with Plastic, that will tout the materials' benefits while also rebutting what the industry considers the huge amount of misinformation about plastics on the Web. Mike Verespej, "Playing Offense," *Plastics News*, June 29, 2009.

the group spent $5.7 million: Lobbying disclosure reports filed with the California secretary of state's office. The 2007–2008 session was also the period during which California's Green Chemistry Initiative was being debated, a measure in which the ACC had a far greater stake. As of the third quarter of 2010, the group reported it had spent nearly $1.2 million on the 2009–2010 legislative session, though not all on bag-related efforts. The same session, the ACC was fighting a proposed ban on bisphenol A. At the federal level, the ACC spent $4.9 million on lobbying in 2008 and more than $8 million in 2009, according to www.opensecrets.org.

164 *polling conducted by the city:* Author interviews with Keith Christman, American Chemistry Council, December 2009, and Dick Lilly, waste-prevention business-area manager, Seattle Public Utilities, and one of the framers of the fee measure, December 2009. Interestingly, at the same time the city council passed the bag tax, it also enacted a ban on takeout food packages made of expanded polystyrene.

The ACC opted not to challenge the polystyrene ban because, as spokesman Keith Christman put it, "It didn't have the same kind of data going into it." Unlike the bag fee, it was popular, so there was no undercurrent of ambivalence to exploit.

The group spent: Mike Verespej, "Seattle Voters Reject 20-Cent Fee," *Plastics News,* August 19, 2009; Marc Ramirez, "Paper or Plastic? Or Neither?" *Seattle Times,* July 21, 2009. Officially there were other partners in the Coalition to Stop the Seattle Bag Tax, including 7-Eleven and the trade group representing independent Washington groceries. But the bulk of the money came from the ACC's deep pockets.

165 *leaving them outspent:* The Green Bag Coalition spent about $98,000 defending the fees, according to filings with the Seattle Ethics and Election Commission.

pulled out all stops: In addition to its expenditures on lobbying and media, the ACC spent at least $15,000 in political donations during the first nine months of 2010. During that same time, Hilex Poly gave $21,700 to various lawmakers and made a $10,000 contribution to the Democratic State Central Committee. ExxonMobil gave nearly $45,000 to individual legislators, plus $10,000 to the California Republican Party. A group called the Good Chemistry PAC, affiliated with Dow Chemical, gave $3,000 in donations to one influential state senator. To be fair, during the same period the California Grocers Association, which backed the ban, gave $22,000 to legislators, $10,000 to the state's Democratic Party, and $32,500 to the state's Republican Party.

the bags could be a breeding ground: The study, done by reputable scientists Chuck Gerba and David Williams, from the University of Arizona, and Ryan Sinclair, from Loma Linda University, found "large numbers of bacteria" in almost all bags tested, including dangerous *E. coli* bacteria in 12 percent. The solution to the problem was simple: regularly wash the bags.

166 *unlike the earlier generation:* That's according to Green Cities California. See the page on local bag ordinances on its website, http://greencitiescalifornia.org/node/2755. The inclusion of paper bags, along with the fact that those city governments are now armed with an environmental impact report, means it will be harder for Steve Joseph to mount successful legal challenges. San Francisco, meanwhile, is considering a measure to broaden its bag ban to include nearly all other types of plastic bags, with the exception of those used for produce and to wrap newspapers. At the same time, several California bag makers are taking steps to create new kinds of plastic carrier bags: ones that are thicker, and therefore less litterable, and that contain recycled plastic.

The ACC may have succeeded: The list of areas with failed ban proposals includes Edmonds, WA; Fairfax, CA; Malibu, CA; Palo Alto, CA; Ft. McMurray, CA; Brownsville, TX; Maui County, HI; Marshall County, IA; and Westport, CT. Los Angeles passed a ban but was sued by Joseph and agreed to delay implementation pending action by the state legislature.

the District of Columbia: Nikita Stewart, "Bill to Charge Consumers for Bags Prompts Debate," *Washington Post,* April 2, 2009; Tim Craig, "D.C. Bag Tax Collects $150,000 in January for River Cleanup," *Washington Post,* March 30, 2010; Sara Murray and Sudeep Reddy, "Capital Takes Bag Tax in Stride," *Wall Street Journal,* September 20, 2010. Nor was the ACC able to block Brownsville, TX, in 2010 from requiring grocers to impose the highest fee yet slapped on bags: one dollar.

a more significant commitment: Author interview with Steve Russell, managing director, American Chemistry Council, plastics division, December 2009. Also Bruce Horovitz, "Makers of Plastic Bags to Use 40 Percent Recycled Content by 2015," *USA Today,* April 21, 2009.

167 *"It is a little too little":* Murray quoted in "Bag Makers Set Recycling Goal for 2015," *Plastics News,* April 27, 2009.

bag recycling above single-digit rates: There's lots of debate over the numbers. The ACC maintains that bag recycling has grown by leaps and bounds over the past several years and that the various existing programs are now capturing about 13 percent of all bags produced. But convincing analyses by *Plastics News* dispute those claims: they show that bag recycling did jump 24 percent in 2005, the first year the ACC began its push, but since then the rates have scarcely risen at all. According to *Plastic News,* the real rate of bag recycling is only about 8 percent. See "New Data Shows Bag, Film Recycling Stall," *Plastics News,* March 22, 2010; Mike Verespej, "Film, Bag Collections on the Rise," *Plastics News,* March 9, 2009.

168 *Whatever the material:* You can get bogged down in a whole other debate about the environmental impacts of various reusable bags. As the plastics industry pointed out, the polypropylene mesh bags that often appear as replacements to T-shirt bags are mostly made in China, have been found to contain heavy metals, and are not easily recycled. Given the choices available in today's world, we'll find that our best options involve tradeoffs.

Robert Cialdini: Author interview with Cialdini, December 2009. See also Robert Cialdini, "Crafting Normative Messages to Protect the Environment," *Current Directions in Psychological Science* 12 (August 2003): 105–9; Noah Goldstein et al., "Room for Improvement: A Social Psychological Approach to Hotel Environmental Conservation Programs," *Cornell Hotel and Restaurant Administration Quarterly* 48 (2007): 145–50.

169 *An independent consultant:* Robert Lilienfeld, "Report on Field Trip to San Francisco to Assess Plastic Bag Ban," September 2008. Accessed at http://use-less-stuff.com/Field-Report-on-San-Francisco-Plastic-Bag-Ban.pdf. Also, author interview with Lilienfeld, November 2008.

San Franciscans are still consuming: The city used as many as eighty-four million paper bags in 2009, according to calculations by writer Joe Eskenazi. To that end, the city's bag warriors are still hoping to break the paper-bag habit; in 2010 Mirkarimi proposed a law requiring grocers to charge ten cents for every paper or compostable bag.

still awash in plastic bags: The city's own 2008 litter survey found the ban had made no dent in the number of plastic bags tumbling along the city streets; if anything, the number of bags at large was slightly up.

Makers of reusable bags: Ellen Gamerman, "An Inconvenient Bag," September 26, 2008.

7. Closing the Loop

171 *Nathaniel Wyeth often called himself:* Glenn Fowler, "N. C. Wyeth, Inventor, Dies at 78; Developed the Plastic Soda Bottle," *New York Times,* July 7, 1990.

A painter need only: Fenichell, *Plastic,* 316.

"I'm in the same field": Fowler, "N. C. Wyeth."

172 *It only took ten thousand tries:* Jack Challoner, ed., *1001 Inventions That Changed the World* (London: Quintessence, 2009), 835.

Wyeth knew that some polymers: The story of his invention is told in Fenichell, *Plastic,* 315–16. Also see a profile of Wyeth written when he was awarded the prestigious Lemelson Award for innovation, "Nathaniel Wyeth," MIT Inventor of the Week archive, August 1998. Accessed at http://web.mit.edu/invent/iow/wyeth.html. The polymer Wyeth opted to use, PET, is made from the combination of an alcohol, ethylene glycol (antifreeze), and an acid, dimethyl terephthalate, and was discovered in 1941 by British chemists Rex Whinfield and James Dickson. They found the molecule made a good textile, and after the war ended, they launched it under the name terylene. Later manufacturers found other ways to play with the fiber to create different varieties of the wrinkle-proof wash-and-wear fabrics that liberated us from the dismal task of ironing. Emsley, *Molecules at an Exhibition,* 134–35.

polyethylene terephthalate: The terephthalate in PET is from a different branch of the phthalate family than the ones used in vinyl that are suspected of causing hormonal disruptions.

safe enough to win approval: The FDA had already rejected Monsanto's bid to make a liquor bottle made of PVC after evidence surfaced about the potential leaching of the carcinogenic monomer vinyl chloride. Next Coca-Cola and Monsanto collaborated on a $100 million effort to develop a plastic soda bottle made of acrylonitrile styrene. But the effort was abandoned in 1977 after Monsanto acknowledged small amounts of acrylonitrile could leach into the beverage, and researchers found that rats fed a steady diet of the polymer could develop tumors and birth defects. See Fenichell, *Plastic,* 315.

About a third of the 224 billion: Container Recycling Institute, "Sales by container type," accessed at http://www.container-recycling.org/facts/all/data/salesbymat.htm.

173 *By 2000, the average American was guzzling:* The exact figures vary from source to source, but data from the USDA's Economic Research Service suggests that in the early 1970s, each American drank about twenty-two to twenty-five gallons of soft drinks a year. In the late 1990s, soda consumption peaked at nearly fifty-two gallons a year. It's since dropped as more people have turned to teas and juices. See Judy Putnam and Shirley Gerrior, "Americans Consuming More Grains and Veg-

etable, Less Saturated Fat," *Food Review* (September–December 1997). Accessed at http://www.ers.usda.gov/publications/foodreview/sep1997/sept97a.pdf.

Wyeth's wonder also enabled: About a third of beverages are now consumed on the go, according to the What's a Bottle Bill fact sheet posted at www.bottlebill.org.

"immediate consumption channels": Jon Mooallem, "The Unintended Consequences of Hyperhydration," *New York Times Magazine,* May 27, 2007.

Would designer water: Water bottles accounted for half, or thirty-six billion, of the PET bottles produced in 2006, according to the Container Recycling Institute. One marketing consultant told the *New York Times* that her research showed the water-bottle habit was less about physical hydration and more about the psychological comfort of carrying around the bottle. "It's like their bangie," she said, meaning a security blanket. Mooallem, "Unintended Consequences."

It was a change that had its roots: Frank Ackerman, *Why Do We Recycle? Markets, Values, and Public Policy* (Washington, DC: Island Press, 1997), 124–35.

174 *No sooner did Coke and Pepsi:* The first recycling of the bottles took place in 1977.
Wellman had been using: Author interview with Dennis Sabourin, director of the National Association for PET Container Resources (NAPCOR), February 2010.

175 *we recycle only about a quarter:* Association of Postconsumer Plastic Recyclers (APR), "2009 United States National Post-Consumer Plastics Bottle Recycling Report." In 2009 the figure was 28 percent, up from 24 percent in 2007.
That's nearly enough polyester: According to NAPCOR, it takes sixty-three twenty-ounce bottles to make a sweater, so it's more than 870 million sweaters.
It's a collection of energy: Container Recycling Institute, "Energy Impacts of Replacing Beverage Containers Wasted in 2005," accessed at http://www.container-recycling.org/facts/datashow.php?file=/issues/zbcwaste/data/energytable.htm&title=Energy%20Impacts%20%20of%20Replacing%20Beverage%20Containers.
In yet another of plastic's paradoxes: The recycling rate for high-density polyethylene, the #2 plastic used in milk jugs and detergent bottles, is slightly higher than for PET — 29 percent in 2008. But in terms of tonnage, far more PET is recycled, according to the EPA, "Municipal Solid Waste Generation, 2008."
We recycle less plastic: Ibid.

176 *people have been recycling and reusing:* Ackerman, *Why Do We Recycle,* 14.
Americans produced relatively little trash: Strasser, *Waste and Want,* 11–13. The book provides a useful history of how Americans have dealt with and disposed of waste.
"trash heated rooms": Ibid., 13.

177 *The pendulum began to swing back:* Strasser, *Waste and Want,* and Ackerman, *Why Do We Recycle,* are good sources for the history of recycling.
the Mobro's plight: Ackerman, *Why Do We Recycle,* 12.

178 *The code is such a poor guide:* The standards-setting body, ATSM, Inc., is developing changes to the code, some of which may involve adding new number categories. See Mike Verespej, "Changes Planned for Resin Identification Codes Include Categories for PC, PLA," *Plastics News,* October 26, 2009.

shaky commitment: Ackerman, *Why Do We Recycle,* 18–19.

Most Americans now have access: The American Chemistry Council estimates eight out of ten Americans have access to recycling, but some experts say that estimate is too high. Susan Collins, director of the Container Recycling Institute, notes that 40 percent of Americans live in multifamily housing that is generally not well served by curbside recycling programs and that *access* could mean someone who lives in a state that has one drop-off recycling center. Author e-mail correspondence with Collins, July 2010.

182 *California recovers nearly three-fourths:* California Department of Resource and Recovery, "Biannual Report of Beverage Container Sales, Returns, Redemptions and Recycling Rates," May 2010.

six times the average of non-bottle-bill states: The average collection rate in non-bottle-bill states is 13.6 percent, according to the Container Recycling Institute. Bottle-bill states recover 71 percent of soda bottles and 35 percent of noncarbonated bottles. See the CRI website, http://www.container-recycling.org/facts/all/data/recrates-depnon-3mats.htm.

San Francisco calculates: Author e-mail correspondence with Robert Reed, spokesman for Recology, May 2010.

it's located in one of the city's poorest: To site the facility there, the city promised residents of the neighborhood would get first shot at the facility's union-wage jobs.

183 *Plastics are a challenge:* Author interview with Steve Alexander, executive director of the Association of Post-Consumer Plastic Recyclers, December 2008.

184 *seventeen-dollar-an-hour:* Author e-mail correspondence with Robert Reed, July 2010.

There's a certain chicken-or-egg quality: That problem prompted Boston businessman Eric Hudson to found Preserve, a company dedicated to making products out of used polypropylene, the #5 plastic, in the hopes of juicing the economics to encourage recycling of the plastic. Its first product was a toothbrush made of used yogurt containers; it could be mailed backed to the company for further recycling. Nothing Wasted, Everything Gained is the slogan on the package. The company recently partnered with Whole Foods on a campaign, called Gimme Five, that encouraged shoppers to bring their used yogurt containers back to the stores for recycling. The first year, the campaign brought in forty-five thousand pounds of used polypropylene, a drop in the resin bucket. Author interview with Preserve spokesperson C. A. Webb, February 2010.

185 *it's been cheaper for San Francisco:* Author interview with Leno Bellomo, commodities manager, Recology, September 2009.

Surprisingly, by some analyses: Author interview with David Allaway, policy analyst in Solid Waste Program of Oregon Department of Environmental Quality, February 2010.

takes about 70 percent: Toland Lam, quoted in Nina Ying Sun, "China's Lam Talks Up Recycling and Change," *Plastics News,* April 7, 2008.

Much of that is composed of: Actually, the bottles have to be shredded before China will accept them, though China is reconsidering that long-standing policy. If it is rescinded, American recyclers worry it could lead to even more PET bottles being exported overseas. Steve Toloken, "China to Accept Whole PET Bottles," *Plastics News,* December 14, 2009.

186 *The program tracked a shipment:* The *Sixty Minutes* episode "The Electronic Wasteland" originally aired in November 2008.

Lam was one of the early entrepreneurs: Author interview with Toland Lam, March 2009.

187 *later ship back:* Not all recyclers are as fastidious. Another recycling plant I visited mixed different plastics together in its recycling process, and the resulting resin was a lower-quality plastic — fit only for low-end products, like flowerpots and coat hangers. Most of that plant's customers were Chinese manufacturers.

From a global perspective: Author interview with Edward Kosior, managing director Nextek Pty Ltd., a recycling specialist who has designed closed-loop operations around the world, January 2010.

188 *China's hunger for those used goods:* Until recently, all those #3 through #7 plastics were sent straight overseas. With the recession, freight rates skyrocketed, and it became more economical for Recology to find local reprocessors who will take those lower-value plastics.

In 2009, American recycling programs: National Association for PET Container Resources, "2009 Report on Postconsumer PET Container Recycling Activity."

China factor is undercutting American recycling: The reliance on China also left recycling programs around the country hugely vulnerable when the world economy went into free fall in the autumn of 2008. The bottom fell out of the commodities market, China stopped buying all used materials, and the whole recycling infrastructure seized up. Prices cratered for various types of scrap materials, and plastics suffered the most. Recycling programs found themselves selling at a loss, turning away hard-to-recycle plastics, leasing warehouse space to store plastics they couldn't move, and, in some drastic cases, actually landfilling plastics. A few municipalities decided to just close their recycling programs altogether. See Philip Sherwell, "Crash in Trash Creates Mountains of Unwanted Recyclables in the U.S.," *UK Telegraph,* December 13, 2008.

Recycling rates have been dropping: America's overall recycling rate of 34 percent in 2008 is down from the overall rate of 41 percent in 2000, and down twenty percentage points from the all-time high of 54 percent in 1992, according to Container Recycling Institute, "Wasting and Recycling Trends, 2008," 4. Accessed at http://www.container-recycling.org/assets/pdfs/reports/2008-BMDA-conclusions.pdf.

"We get people to do it": Chase Willett, analyst for Chemical Market Associates, Inc., speaking at Plastics Recycling Conference, Austin, Texas, March 2010.

189 *To try to entice more people:* Peter Schworm, "Recycling Efforts Fail to Change Old Habits," *Boston Globe,* March 14, 2010. The problems of single-stream systems are examined in Clarissa Morawski, "Understanding Economic and Envi-

ronmental Impacts of Single-Stream Collection Systems," a report done for the Container Recycling Institute, December 2009.

One recycling expert told me: Author interview with Patty Moore, Moore Consultants, December 2008.

Closed-loop systems: The upstream environmental benefits of closing the loop is considered to be ten to twenty times greater than those gained by downcycling or disposing of a product. Morawski, "Understanding Economic," 8.

190 *supporter of bottle bills:* The first bottle bill was passed in Oregon in 1971. Over the next fifteen years, ten more states followed suit, and then the push for bottle-bill legislation stalled. There were eleven bottle-bill states until 2010, when Delaware repealed its twenty-eight-year-old five-cent deposit. State lawmakers said the measure didn't lead to many bottle returns because most stores refused to take the containers back. So the legislature replaced it with a nonrefundable four-cent fee, which is supposed to provide start-up funds for waste haulers to set up curbside recycling programs. Mike Verespej, "Delaware Replaces Bottle Deposits with Controversial Fee," *Plastics News,* May 17, 2010.

Plastic Pollution Texas: The group was started by Mike Garvey, a water-pollution activist in Houston. Author interviews with Mary Wood and Patsy Gillham, March 2010.

the specifics of deposit laws vary: Author interview with Collins; also "What Is a Bottle Bill," http://www.bottlebill.org/about/whatis.htm.

191 *bottle-bill states have at least twice the recovery:* Container Recycling Institute website.

gets more than 90 percent: Ibid. Beverage container litter has dropped by anywhere from 69 to 84 percent in bottle-bill states.

192 *Aside from Hawaii:* Author interview with Collins and her predecessor Betty McLaughlin, October 2007. See also Mooallem, "Unintended Consequences," on the fights over broadening container-deposit laws to include bottled water.

Howard Rappaport: Author interview with Rappaport.

193 *for every pound of trash:* Brenda Platt et al., *Stop Trashing the Climate* (Washington, DC: Institute for Local Self-Reliance, 2008), 19.

"The recycling movement has missed the forest": Author interview with Bill Sheehan, executive director, Product Policy Institute, February 2010.

194 *In 1970, the average American:* EPA, "Municipal Solid Waste," 2008, 9.

more plastic packaging: The Grassroots Recycling Network report "Wasting and Recycling in the U.S. 2000" indicates that between 1990 and 1997, plastic packaging grew five times faster by weight than plastic recovered for recycling; cited in Jim Motavalli, "Zero Waste," *E* magazine (March/April 2001).

"a rite of atonement": John Tierney, "Recycling Is Garbage," *New York Times Magazine,* June 30, 1996.

"you make it": Lyle Clarke, vice president policy and programs, Stewardship Ontario, which manages one of the province's EPR programs, speaking at Plastics Recycling Conference, Austin, Texas, March 2010.

Sheehan maintains this is perfectly logical: The full argument is laid out in Helen Spiegelman and Bill Sheehan, "Unintended Consequences: Municipal Solid Waste Management and the Throwaway Society" (Athens, GA: Product Policy Institute, March 2005). See also Melinda Burns, "The Smoldering Trash Revolt," *Miller-McCune* magazine, January 21, 2010.

195 *the first explicit EPR law:* Information on the German system comes from Imhoff, *Paper or Plastic,* 46–53; Clean Production Action, "Summary of Germany's Packaging Takeback Law," September 2003; accessed at www.cleanproduction.org/library/EPR_dvd/DualesSystemDeutsch_REVISEDoverview.pdf. A good overview of the program is Betty Fishbein, "EPR: What Does It Mean? Where Is It Headed?" *Pollution Prevention Review* 8 (1998): 43–55; accessed at www.informinc.org/eprppr.phpP2.

The law has accomplished: "Profits Warning: Why Germany's Green Dot Is Selling Up," *Let's Recycle,* November 25, 2004, accessed at http://www.letsrecycle.com/do/ecco.py/view_item?listid=38&listcatid=218&listitemid=2056&section=.

196 *Europe now diverts more than half:* PlasticsEurope, "Compelling Facts About Plastic, 2009"; accessed at http://www.plasticseurope.org/Documents/Document/20100225141556-Brochure_UK_FactsFigures_2009_22sept_6_Final-20090930-001-EN-v1.pdf.

technology known as waste-to-energy: Overall, about 30 percent of the plastic diverted from landfills in Europe is burned for energy. Ibid. See also Elisabeth Rosenthal, "Europe Finds Cleaner Source of Fuel in Trash Incinerators as U.S. Sits Back," *New York Times,* April 13, 2010. Critics contend that the plants may generate greenhouse gases, produce toxic ash, and encourage increased production of garbage, since they rely on it for their feedstock.

197 *refillable bottles:* Author interview with Collins.

Green Dot law cut packaging: Fishbein, "EPR." A more recent report, prepared for the California Department of Conservation, found that since 2000, the amount of packaging waste produced has leveled out at between 15.1 and 15.5 million metric tons, suggesting the policies have succeeded in severing the connection between economic growth and increasing waste; see R3 Consulting Group and Clarissa Morawski, "Evaluating End-of-Life Beverage Container Management Systems for California," California Department of Conservation, May 2009.

More than thirty countries: California Ocean Protection Council, "An Implementation Strategy for the California Ocean Protection Council Resolution to Reduce and Prevent Ocean Litter," November 20, 2008, 11.

In California, Vermont, Oregon: Product Policy Institute website; California Product Stewardship Council website.

198 *Nature, they pointed out:* William McDonough and Michael Braungart, *Cradle to Cradle: Remaking the Way We Make Things* (New York: North Point Press, 2002), 92.

I attended one of its meetings: Sustainable Packaging Coalition quarterly meeting, San Francisco, April 2008.

199 *Take the example of the juice boxes:* Author interview with Ann Johnson, director
of Sustainable Packaging Coalition, March 2008.

Coca-Cola has been concerned: Information on Coca-Cola's various measures
comes from author interview with Scott Vitters, director of sustainable packaging, environmental, and water resources, Coca-Cola, March 2010, and author
e-mail correspondence with him, August 2010. Also see Coca-Cola, Our Commitment to Environmental Stewardship press release; Marc Gunther, "Coca-Cola's Green Crusader," *Fortune*, April 28, 2008; Marc Gunther, "Coca-Cola's New
PlantBottle Sows Path to Greener Packaging," Greenbiz.com, December 1, 2009.

200 *lightest PET soda bottle on the market:* Amy Galland, "Waste and Opportunity:
U.S. Beverage Container Scorecard and Report, 2008." The report, written for the
watchdog group As You Sow, gave Coca-Cola an overall grade of C on a scorecard
that assessed a range of criteria including source reduction, recycling, and the use
of recycled content. Tellingly, that was the highest score of any beverage manufacturer.

Dasani water bottles: Betsy McKay, "Message in the Drink Bottle: Recycle," *Wall
Street Journal,* August 30, 2007.

I LOHAS: Ariel Schwartz, "Coca-Cola Japan Sells Easy-Crush Water Bottles to
Save Plastic, But Is It Greenwashing?" *Fast Company,* June 9, 2009; accessed at
http://www.fastcompany.com/blog/ariel-schwartz/sustainability/coca-cola-
japan-selling-easy-crush-water-bottles-save-plastic-it-?#.

201 *Vitters said one reason:* Author e-mail correspondence with Vitters, August 2010.

factory will have the capacity: Mike Verespej, "Coke Planning U.S. PET Recycling
Plant," *Plastics News,* August 31, 2007; editorial, "Coke's PET Pledge: Real Progress or PR?" *Plastics News,* September 24, 2007.

not to much more than 50 percent: Author interview with Kosior.

8. The Meaning of Green

203 *One evening in 1951:* Caitlin McDevitt, "Plastics Flashback: A Visual History of
the Credit Card," *Big Money,* May 28, 2009; accessed at http://www.thebigmoney
.com/slideshow/plastic-flashback#.

204 *"better withstand day-to-day use":* Ibid. Around the same time, a Cleveland company began producing plastic charge cards, promoting them with the suggestion
that the cards would serve as "billfold billboards" for the businesses that issued
them. The idea quickly caught on with the major gas companies as well as with
stores and banks; see "Credit Cards in Plastics," *Modern Plastics,* November 1957.

"She had a whole purse full": Oxford English Dictionary; author e-mail correspondence with Geoffrey Nunberg, April 2010.

"A plastic card is a physical device": The phrase appeared on the website of Teraco,
one of the leading manufacturers of plastic cards.

Four out of five Americans: "The Survey of Consumer Payment Choice," Federal

Reserve Bank of Boston, January 2010; accessed through www.creditcard.com.

cards are also increasingly the stand-ins: Tracie Rozhon, "The Weary Holiday Shopper Is Giving Plastic This Season," *New York Times,* December 9, 2002.

ten billion are now created: Cindy Waxer, "Eco-friendly Initiatives Focus on Gift Cards," Creditcards.com; accessed at http://www.creditcards.com/credit-card-news/eco-friendly-green-gift-cards-plastic-1273.php.

205 *Card issuers have played on status consciousness:* McDevitt, "Plastics Flashback."

206 *"We are encouraged":* Author e-mail correspondence with Mai Lee, media relations, Discover Card Services, April 2010.

Card manufacturers like PVC: Author interview with John Kiekhaefer, development manager, Perfect Plastic Cards, May 2010.

there are more than 1.5 billion: Laura Shin, "Making Credit Cards Landfill Friendly," *New York Times* Green blog, http://green.blogs.nytimes.com/2009/02/23/making-credit-cards-landfill-friendly/.

more than seventy-five million: That's according to estimates by Rodd Gilbert, owner of Earthworks Systems, a Solon, Ohio–based company that collects and recycles used plastic cards. Author interview with Gilbert, April 2010.

The thought of all those plastic cards: Author interview with Paul Kappus, April 2010.

207 *"That's a load of hooey":* Author interview with Tim Greiner, partner, Pure Strategies, April 2010.

208 *Production of biobased polymers:* Li Shen et al., "Product Overview." According to this report on the future of bioplastics, the global average annual growth rate was 38 percent between 2003 and 2007 and as high as 48 percent in Europe. Other forecasts project growth rates for bioplastics of 15 to 20 percent for 2011, and anywhere from 12 to 40 percent annually in following years, depending on how quickly new resins and new markets develop. See Mike Verespej, "Despite New Feedstocks, Resins Will Be Resins," *Plastics News,* August 16, 2010.

a drop in the resin bucket: Author interview with Ramani Narayan, professor of chemistry, Michigan State University at Lansing, May 2010. Worldwide, 360,000 tons of bioplastics were produced in 2007, accounting for just 0.3 percent of global plastics production. Jon Evans, "Bioplastics Get Growing," *Plastics Engineering* (February 2010): 16

bioplastics could one day replace: Li Shen et al., "Product Overview," 2.

"Our whole industry agrees": Mauro Gregorio, head of alternative feedstocks for Dow, quoted in Joshua Schneyer, "Brazil's 'Organic' Plastics," *BusinessWeek,* June 24, 2008.

209 *agricultural interests competed:* Nonny de la Pena, "Bioplastics Lifts Garbage Out of the Trash Heap," *New York Times,* June 19, 2007.

Henry Ford: The story of Ford's interest in soybeans is told in Meikle, *American Plastic,* 155–56, and in Geiser, *Materials Matter,* 260–61. The *Time* quote is from Meikle, *American Plastic,* 156. In the 1946 movie *It's a Wonderful Life,* George

Bailey's friend Sam tries to get him to invest in soy plastics. George declines and remains a small-town banker in Bedford Falls while his buddy makes a bundle molding soy-acrylic bubbles for airplanes (ibid., 159).

210 *Oil was inexpensive:* de la Pena, "Bioplastics."

Only a few plant-based polymers: In the 1980s and '90, in response to concerns about plastic piling up in landfills, manufacturers created varieties of pseudo-biopolymers — conventional plastics with plant starches blended in that would theoretically biodegrade. What actually happened was that only the starch components broke down, leaving the petro-based polymers behind. Not only were the materials a nonsolution to the waste issues posed by plastics but the products made of them tended to rip or break, which had the effect of tarnishing the whole field of bioplastics for a while. It was reminiscent of how the reputation of early thermoplastics had been damaged by the shoddy products made in the years following World War II.

"Carbon is carbon": Kerry Dolan, "Revving Up Nature's Engines," *Forbes,* July 24, 2006.

Brazil's petrochemical giant: Frank Esposito, "Biopolymers Building Muscle in Market," *Plastics News,* March 22, 2010.

Sustainable Monopoly: Rosalie Morales et al., "The Brazilian Bioplastics Revolution," a paper published on Wharton Business School's Knowledge@Wharton website; accessed at http://www.wharton.universia.net/index.cfm?fa=viewArticle&id=1704&language=english Wharton study.

hub for cane-based plastics: Procter and Gamble already uses cane-based polyethylene in packaging for several of its products, including some Max Factor and CoverGirl cosmetics. Toyota plans to use cane-based PET for the interior of its cars.

soy-based polyurethane cushions: The company is also using seat fabrics made from recycled yarn in the Ford Escape and the Escape Hybrid. And in the 2010 Ford Flex, the storage bins are made from wheat-straw-reinforced plastic. Ford's goal is to eventually make cars in which all the plastic components come from compostable materials. Rhoda Miel, "Natural Fiber Use in Auto Parts Expands," *Plastics News,* December 4, 2009; Candace Lombardi, "Our Cars Are 85 Percent Recyclable, Ford says," *CNET News,* April 22, 2010.

211 *"Just because it's biobased":* Author interview with Mark Rossi, October 2009.

212 *two different scorecards:* Versions of the scorecards can be accessed online through the websites of Clean Production Action, cleanproduction.org, and the Sustainable Biomaterials Collaborative, www.sustainablebiomaterials.org.

213 *NatureWorks makes a corn-based polymer:* PLA was actually first synthesized over a hundred and fifty years ago, but no application for it was found until the 1960s, when it became apparent it could be useful in medicine, since the material dissolved harmlessly inside the body. Not until the late 1980s did various companies, including DuPont, Coors, and Cargill, start exploring how to scale up production

to develop it as a commodity plastic; see Li Shen et al., "Product Overview," 57. Background on NatureWorks and PLA comes from author interview with Steve Davies, NatureWorks spokesman, April 2010, and from Elizabeth Royte, "Corn Plastic to the Rescue," *Smithsonian,* August 2006.

214 *change can be tricky:* Suzanne Vranica, "Snack Attack: Chip Eaters Make Noise About a Crunchy Bag," *Wall Street Journal,* October 5, 2010.

215 *Among its shortcomings:* Because there are limits to PLA's versatility, a company called Cereplast is building a healthy business creating hybrids of PLA and conventional petro-plastics that extend the biopolymer's range of attributes.

biotech company, Metabolix: Background on Metabolix comes from author interviews with spokesman Brian Igoe and company founder Oliver Peoples, July 2009. Also see Mara Der Hovanesian, "I Have Just One Word for You: Bioplastics," *BusinessWeek,* June 19, 2008.

216 *The plant would be harvested:* In 2010 Peoples said the company was "one to two years from field trials with our commercial crops," which would be the next step toward commercialization.

Monsanto reported that it had succeeded: Evans, "Bioplastics Get Growing," 17.

217 *Narayan . . . had a single answer:* Author interview with Narayan. Some critics have questioned if the benefits are really that great, at least in the case of Metabolix. A former Metabolix scientist argued that the energy required to convert corn sugar to PHA exceeds that needed to make conventional polyethylene. The same would be true to extract PHA from switchgrass or some other nonfood crop. And he contended that because it's compostable, it can release methane, whereas nonbiodegradable petro-plastics will act as long-term carbon sequesterers. When you consider all the factors, Dartmouth College engineer Tillman Gerngross said, Mirel "is unsustainable." Author interview with Gerngross, April 2010; Gerngross, "How Green Are Green Plastics?" *Scientific American,* August 2000. Narayan maintained Gerngross's analysis didn't give adequate weight to the initial carbon savings gained by the use of renewable feedstocks.

218 *PLA produces just 1.3 kilograms:* E.T.H. Vink et al., "The Eco-profiles for Current and Near-future NatureWorks Polylactide (PLA) Production," *Industrial Biotechnology* 3 (2007): 58–81.

genetically modified crops: The Sustainable Biomaterials Collaborative—a network of environmental groups and businesses interested in bioplastics—opposes the use of biopolymers made from genetically modified feedstocks. The group is especially concerned about GM switchgrass, since that's an open-pollinating crop that could threaten nearby natural switchgrass. Author e-mail correspondence with SBC cochair Brenda Platt, August 2010.

such sources add up to 350 million: Geiser, *Materials Matter,* 331.

219 *Criddle is working with methane-eating:* Author interview with Craig Criddle, Stanford University, May 2010.

Cornell University chemist: Stacey Shackford, "Ithaca Plastics Company Gets

$18.4 Million Federal Grant," *Ithaca Journal,* July 22, 2010. Coates is commercializing his work through a private company that has been awarded more than twenty million dollars in federal grants.

220 *Green chemistry:* Since the early 1960s, scientists have become increasingly concerned with the environmental implications of traditional chemistry practices. But green chemistry as a field didn't really get going until the late 1980s, when two leading scientists, John Warner and Paul Anastas, laid out a set of guiding principles for the field. Those principles started with the declaration that "it is better to prevent waste than to treat or clean up waste after it is formed" and went on to emphasize more environmentally friendly practices, such as reducing the use of solvents, separation agents, and other auxiliary substances whenever possible; using renewable raw materials where technically and economically possible; designing products so they don't persist in the environment and do break down into innocuous byproducts; and choosing substances and processes that have a minimal chance of causing chemical accidents. J. A. Linthorst, "An Overview: Origins and Development of Green Chemistry," *Foundations of Chemistry* 12 (2010): 55–68.

requires any manufacturer: The list doesn't explicitly bar the use of bisphenol A or phthalates, but as PLA gets picked up for more applications, Davies said the company is considering whether it needs to spell that out.

biodegradable *in this context:* Author interviews with Narayan and Steve Mojo, executive director of the Sustainable Products Institute, April and May 2010. See also Ramani Narayan, "Misleading Claims and Misuse of Standards Continues to Proliferate in the Nascent BioPlastics Industry Space," *Bioplastics* magazine, January 2, 2010; Brenda Platt, "Biodegradable Plastics: True or False, Good or Bad?" accessed at http://www.sustainableplastics.org/spotlight/biodegradable-plastics-true-or-false-good-or-bad. For a product to be called *biodegradable,* its manufacturer needs to provide specifics as to how long and under what conditions it will break down. Those issues are at the crux of industry standards for biodegradability.

Narayan has criticized my purportedly biodegradable: Author interview with Narayan.

"oxo-biodegradable": Author interview with Mojo. One study of oxo-biodegradable bags sponsored by the California Integrated Waste Management Board found no evidence of biodegradation, which led the State of California to pass a law restricting the use of the terms *compostable biodegradable, degradable,* and *marine degradable* on plastic bags.

221 *fossil-fuel-based plastics that will biodegrade:* For instance, BASF's Ecoflex plastic is a petroleum-based polymer that's biodegradable and compostable. It's used to make compostable bags.

U.S. Navy is exploring: Any plans to purchase Mirel food ware are on hold, however, because the United States is bound by the Marpol Treaty, which bars the dumping of waste at sea. There are efforts under way to revise the treaty to "support plastics that are marine-degradable," according to author e-mail correspondence with Metabolix spokesman Brian Ruby, August 2010.

only about two hundred: Rhodes Yepsen, "U.S. Residential Food Waste Collection and Composting," *BioCycle* (December 2009): 35.

222 *plastic is creating a mini-crisis:* According to *Time* magazine, some composting centers have a blanket policy of discarding all plastic. "I direct pickers to take out plastic, which they can't distinguish from bioplastic," said Will Bakx, co-owner of and soil scientist at Sonoma Compost, a composting facility in Petaluma, California. Kristina Dell, "The Promise and Pitfalls of Bioplastic," *Time* magazine, May 3, 2010.

"Plastic is forever": Quoted in Meikle, *American Plastic,* 9.

It's a myth and a misplaced hope: The Federal Trade Commission considers it illegal for plastic products to claim they can biodegrade in a landfill. "The trouble is they haven't sent enough people to prison for doing so," Mojo observed. New FTC guidelines, however, promise to make it harder for companies to make unsubstantiated green claims. An ad proclaiming a product is ecofriendly will have to be backed up with evidence and specific supporting phrases, such as "based on the ability to recycle." Jack Neff, "FTC Goes After Broad Environmental Claims," *Advertising Age,* October 6, 2010.

223 *Americans throw away:* EPA, "Municipal Solid Waste, 2008."

the one chosen by HSBC: "DiCaprio Promotes Green Credit Cards," *Huffington Post,* March 24, 2008. Also described on HSBC's website; see http://www.hsbc .com.hk/1/2/cr/environment/projects/green_credit_card.

224 *recent report in the New York Times:* Penelope Green, "Biodegradable Home Lines, Ready to Rot," *New York Times,* May 8, 2008.

225 *"inflationary culture":* Meikle, *American Plastic,* 176.

As Terry tells the story: Author interview with Beth Terry, December 2008, April 2010. Also see her blog, www.fakeplasticfish.com.

226 *Yet another of the ubiquitous chemical's uses:* Janet Raloff, "Concerned About BPA: Check Your Receipts," *Science News,* October 7, 2009. Accessed at http:// www.sciencenews.org/view/generic/id/48084/title/Science_%2B_the_Public__ Concerned_about_BPA_Check_your_receipts.

230 *reminded me of something:* Author interview with Lilienfeld.

232 *annual global plastics production:* Li Shen et al., "Product Overview," ii.

Epilogue: A Bridge

233 *a spokeswoman for the state forest:* Author interview with Terry Schmidt, customer service representative, Wharton State Forest, May 2010.

Nearly one million used milk jugs: Author interview with Jim Kerstein, founder and CEO of Axion International, May 2010. For more information on the company's bridges, see Sergio Bichao, "Lightweight—But Strong—Plastic Used by Army to Build Bridge Has Origins in Rutgers Labs," *Courier News*, October 9, 2009; accessed at http://www.mycentraljersey.com/article/20091009/NEWS/91009050.

234 *"Others build strong bridges"*: A video of the bridge's dedication can be seen at http://www.youtube.com/watch?v=ohE-ymdio44.

These bridges are slated: Author interview with Marc Green, president of Axion International, May 2010.

The deteriorating wooden bridge: Author interview with Schmidt.

Selected Bibliography

Ackerman, Frank. *Poisoned for Pennies: The Economics of Toxics and Precaution.* Washington, DC: Island Press, 2008.

——. *Why Do We Recycle?: Markets, Values, and Public Policy.* Washington, DC: Island Press, 1997.

Andrady, Anthony. *Plastics and the Environment.* Hoboken, NJ: John Wiley and Sons, 2003.

Baker, Nena. *The Body Toxic: How the Chemistry of Everyday Things Threatens Our Health and Well-Being.* New York: North Point Press, 2008.

Chang, Leslie. *Factory Girls: From Village to City in a Changing China.* New York: Spiegel and Grau, 2009.

Clarke, Alison J. *Tupperware: The Promise of Plastic in 1950s America.* Washington, DC: Smithsonian Institution Press, 1999.

Coe, J. M., and D. B. Rogers, eds. *Marine Debris — Sources, Impacts, and Solutions.* New York: Springer-Verlag, 1997.

Colborn, Theo, Dianne Dumanoski, and John Peterson Myers. *Our Stolen Future: Are We Threatening Our Fertility, Intelligence, and Survival? A Scientific Detective Story.* New York: Dutton, 1996.

DiNoto, Andrea. *Art Plastic: Designed for Living.* New York: Abbeville Press, 1987.

Doyle, Bernard. *Comb Making in America: An Account of the Origin and Development of the Industry for Which Leominster Has Become Famous.* Privately printed, 1925.

Doyle, Jack. *Trespass Against Us: Dow Chemical and the Toxic Century.* Monroe, ME: Common Courage Press, 2004.

DuBois, J. Harry. *Plastics History, U.S.A.* Boston: Cahners Books, 1972.

Ebbesmeyer, Curtis, and Eric Scigliano. *Flotsametrics and the Floating World: How One Man's Obsession with Runaway Sneakers and Rubber Ducks Revolutionized Ocean Science.* New York: HarperCollins, 2009.

Emsley, John. *Molecules at an Exhibition: The Science of Everyday Life.* Oxford: Oxford University Press, 1998.

Fenichell, Stephen. *Plastic: The Making of a Synthetic Century.* New York: HarperCollins, 1996.

Fiell, Charlotte, and Peter Fiell. *1000 Chairs.* Cologne, Ger.: Taschen, 2000.

Friedel, Robert. *Pioneer Plastic: The Making and Selling of Celluloid.* Madison: University of Wisconsin Press, 1983.

Geiser, Kenneth. *Materials Matter: Toward a Sustainable Materials Policy.* Cambridge, MA: MIT Press, 2001.

Hammond, Ray. *The World in 2030.* Itxaropena, Spain: Editions Yago, 2007.

Hill, John W., and Doris Kolb. *Chemistry for Changing Times.* Saddle River, NJ: Prentice Hall, 1998.

Hine, Thomas. *Populuxe.* New York: Knopf, 1986.

——. *The Total Package: The Evolution and Secret Meaning of Boxes, Bottles, Cans, and Tubes.* Boston: Little, Brown, 1995.

Imhoff, Daniel. *Paper or Plastic: Searching for Solutions to an Overpackaged World.* San Francisco: Sierra Club Books, 2005.

Katz, Sylvia. *Plastics: Common Objects, Classic Designs.* New York: Harry N. Abrams, 1984.

Lauer, Keith, and Julie Robinson. *Celluloid: Collector's Reference and Value Guide.* Paducah, KY: Collector Books, 1999.

Mark, Herman F. *Giant Molecules.* New York: Time-Life Books, 1966.

Markowitz, Gerald, and David Rosner. *Deceit and Denial: The Deadly Politics of Industrial Pollution.* Berkeley: University of California Press, 2002.

McDonough, William, and Michael Braungart. *Cradle to Cradle: Remaking the Way We Make Things.* New York: North Point Press, 2002.

Meikle, Jeffrey. *American Plastic: A Cultural History.* New Brunswick, NJ: Rutgers University Press, 1997.

Mossman, Susan, ed. *Early Plastics: Perspectives, 1850–1950.* London: Leicester University Press, 2000.

Newman, Thelma, R. *Plastics as Design Form.* Philadelphia: Chilton Book Co., 1972.

Odian, George. *Principles of Polymerization.* 4th ed. Hoboken, NJ: John Wiley and Sons, 2004.

Rogers, Heather. *Gone Tomorrow: The Hidden Life of Garbage.* New York: New Press, 2005.

Rosato, Dominick, William Fallon, and Donald Rosato. *Markets for Plastics.* New York: Van Nostrand Reinhold, 1969.

Rosato, Donald, Marlene Rosato, and Dominick Rosato. *Concise Encyclopedia of Plastics.* Boston: Kluwer Academic Publishers, 2000.

Royte, Elizabeth. *Bottlemania: How Water Went on Sale and Why We Bought It.* New York: Bloomsbury, 2008.

———. *Garbage Land: On the Secret Trail of Trash.* Boston: Little, Brown, 2005.

Schapiro, Mark. *Exposed: The Toxic Chemistry of Everyday Products and What's at Stake for American Power.* White River Junction, VT: Chelsea Green Publishing, 2007.

Sparke, Penny, ed. *The Plastics Age: From Bakelite to Beanbags and Beyond.* Woodstock, NY: Overlook Press, 1993.

Strasser, Susan. *Satisfaction Guaranteed: The Making of the American Mass Market.* New York: Pantheon Books, 1989.

———. *Waste and Want: A Social History of Trash.* New York: Metropolitan Books, 1999.

von Vegesack, Alexander, and Mathias Remmele, eds. *Verner Panton: The Collected Works.* Weil am Rhein, Ger.: Vitra Design Museum, 2000.

Walsh, Tim. *Wham-O Super-Book: Celebrating Sixty Years Inside the Fun Factory.* San Francisco: Chronicle Books, 2008.

Wargo, John. *Green Intelligence: Creating Environments That Protect Human Health.* New Haven, CT: Yale University Press, 2009.

Weisman, Alan. *The World Without Us.* New York: Thomas Dunne Books, 2007.

Whitehead, Don. *The Dow Story: The History of the Dow Chemical Company.* New York: McGraw-Hill, 1968.

Yarsley, V. E., and E. G. Couzens. *Plastics.* Harmondsworth, UK: Penguin Books, 1941.

Index